Adaptive Mesh Refinement for Time-Domain Numerical Electromagnetics

Adaptive Mesh Refinement for Time-Domain Numerical Electromagnetics

Costas D. Sarris

ISBN: 978-3-031-00567-1   paperback

ISBN: 978-3-031-00567-1   paperback

ISBN: 978-3-031-01695-0   ebook

ISBN: 978-3-031-01695-0   ebook

DOI: 10.1007/978-3-031-01695-0

A Publication in the Springer series
SYNTHESIS LECTURES ON COMPUTATIONAL ELECTROMAGNETICS #11
Series Editor: Constantine A. Balanis, Arizona State University

ISSN: 1932-1252 (Print)
ISSN: 1932-1716 (Electronic)

First Edition
10 9 8 7 6 5 4 3 2 1

Adaptive Mesh Refinement for Time-Domain Numerical Electromagnetics

Costas D. Sarris
University of Toronto
Ontario, Canada

SYNTHESIS LECTURES ON COMPUTATIONAL ELECTROMAGNETICS #11

ABSTRACT

This monograph is a comprehensive presentation of state-of-the-art methodologies that can dramatically enhance the efficiency of the finite-difference time-domain (FDTD) technique, the most popular electromagnetic field solver of the time-domain form of Maxwell's equations. These methodologies are aimed at optimally tailoring the computational resources needed for the wideband simulation of microwave and optical structures to their geometry, as well as the nature of the field solutions they support. That is achieved by the development of *robust* "adaptive meshing" approaches, which amount to varying the total number of unknown field quantities in the course of the simulation to adapt to temporally or spatially localized field features. While mesh adaptation is an extremely desirable FDTD feature, known to reduce simulation times by orders of magnitude, it is not always robust. The specific techniques presented in this book *are characterized by stability and robustness*. Therefore, they are excellent computer analysis and design (CAD) tools.

The book starts by introducing the FDTD technique, along with challenges related to its application to the analysis of real-life microwave and optical structures. It then proceeds to developing an adaptive mesh refinement method based on the use of multiresolution analysis and, more specifically, the Haar wavelet basis. Furthermore, a new method to embed a moving adaptive mesh in FDTD, the dynamically adaptive mesh refinement (AMR) FDTD technique, is introduced and explained in detail. To highlight the properties of the theoretical tools developed in the text, a number of applications are presented, including:

- Microwave integrated circuits (microstrip filters, couplers, spiral inductors, cavities).
- Optical power splitters, Y-junctions, and couplers
- Optical ring resonators
- Nonlinear optical waveguides.

Building on first principles of time-domain electromagnetic simulations, this book presents advanced concepts and cutting-edge modeling techniques in an intuitive way for programmers, engineers, and graduate students.

It is designed to provide a solid reference for highly efficient time-domain solvers, employed in a wide range of exciting applications in microwave/millimeter-wave and optical engineering.

KEYWORDS

Computational electromagentics, finite difference time domain (FDTD), multiresolution time domain (MRTD), adaptive mesh refinement, microwave simulation, optical waveguides.

Contents

Acknowledgments

First and foremost, I would like to express my gratitude to Prof. Constantine Balanis, for asking me to contribute to this series, and to Joel Claypool for his support to this project.

The research included in the part of this monograph that refers to the MRTD technique was conducted at the University of Michigan, Ann Arbor, under the inspiring guidance of Prof. Linda Katehi, to whom the author is ever indebted for her mentorship and example of achievement through excellence. Financially, this work was supported by the Army Research Office (and Dr. James F. Harvey should be personally acknowledged for his continued interest) and the U.S. Army CECOM (with Dr. Barry Perlman deserving a special acknowledgment as well).

The work on the dynamically adaptive mesh refinement finite-difference time-domain (AMR-FDTD) technique is the result of a productive collaboration of the author with Dr. Yaxun Liu at the Edward S. Rogers Sr. Department of Electrical and Computer Engineering, University of Toronto. It was primarily supported by the Natural Sciences and Engineering Research Council of Canada (NSERC) through a Discovery Grant to the author. Dr. Liu was partially supported by the Ontario Centers of Excellence through a Collaborative Research Grant and NSERC through a Collaborative Research and Development (CRD) Grant. All this generous support is gratefully acknowledged.

Costas D. Sarris, *Toronto, Ontario, September 2006*

List of Tables

List of Figures

CHAPTER 1

Introduction

Recent advances in solid-state device technology have enabled the development of planar monolithic microwave integrated circuits (MMICs) in multilayer, densely integrated architectures. These provide low-cost, lightweight and reproducible alternatives to conventional, low-loss but bulky components. In addition, circuit miniaturization allows for the implementation of system-on-a-chip concepts, that serve the need of the wireless industry for light and small transceiver structures well.

The next steps beyond the current state-of-the-art include the monolithic integration of silicon/silicon germanium (SiGe) circuits, microelectromechanical switch (MEMS) devices, micromachined components (e.g., filters, resonators, multiplexers), and digital CMOS signal processing units, for high power/low-loss, tactical and commercial communication systems.

As a result, approximate circuit design approaches, which are based on representing multielement structures as an assembly of cascaded, individually functioning blocks, do not constitute a reliable simulation tool for cutting-edge applications, due to their failure to account for the inevitable electromagnetic coupling of closely spaced components. Yet, the latter produces parasitic effects such as crosstalk and signal distortion, that generate signal integrity issues with an adverse impact on the overall performance of wireless communication systems. The profound need to model and eventually mitigate such effects fuels the research on *full-wave techniques* of computational electromagnetics (CEM). Methods that are referred to as such, proceed to the formulation of a solution to a given boundary value electromagnetic problem, with no assumptions on the shape of the wavefronts involved. On the other hand, high-frequency asymptotics assume the reduction of wavefronts to rays, facilitating the application of optics principles to problems of scattering from electrically large, unconventionally shaped objects [1].

In essence, numerical schemes are developed by formulating an operator problem and subsequently pursuing its solution by computational (as opposed to analytical) means. The question of the operator itself being either an integral or a differential one, classifies accordingly the related scheme. Electromagnetic problems are cast in the form of an integral equation by use of a Green function, which assumes in general the form of a dyad $\overline{\overline{G}}(\bar{r}, \bar{r}')$. Despite the fact that special forms of integral equations can be treated by analytical techniques, such as, among others, the Wiener-Hopf method [2], the dominant integral equation solver is arguably the numerical method of moments [3].

A Green function incorporates an intuitively rich description of the physics of a given problem. In addition, it inherently accounts for Sommerfeld's radiation condition [4], rendering the use of absorbing boundaries unnecessary. However, the use of Green dyadics becomes extremely complex, if tractable at all, in realistic cases, naturally giving way to differential methods, that trade versatility with computational effort. In fact, techniques such as the finite element method (FEM, [5]), the finite difference time-domain (FDTD, [6]), or the transmission line matrix (TLM, [7]) method that solve partial differential equations in a mesh, translate into large-scale algebraic systems. Nevertheless, recent advances in high-performance and parallel computing have made these methods a feasible alternative, thus prompting an ever-increasing research interest in those.

This monograph focuses on the solution of Maxwell's equations in the time-domain. To this end, a broadband impulse excitation, such as a Gaussian pulse, is injected as an initial condition to the computational domain. Then, a marching-in-time procedure is followed to calculate the evolution of its wavefront, until the steady state is reached. On the contrary, frequency-domain techniques extract field information at a given frequency point. Clearly, the latter are preferred to the former, when narrowband simulations are pursued. In addition, frequency sweep techniques, such as the asymptotic waveform evaluation (AWE, [8]), have enhanced the capability of frequency-domain techniques to produce broadband results in certain cases, suffering though from general issues of accuracy and efficiency.

Time-domain differential methods are becoming increasingly popular among the electromagnetic community as a consequence of their versatility and their ability to provide simulation results that are intuitively meaningful to circuit designers and microwave engineers. In particular, the finite-difference time-domain (FDTD) technique offers a mathematically straightforward analysis method, suitable for arbitrary electromagnetic geometries. However, since Yee's scheme [6] is second-order accurate only, and sensitive to numerical dispersion, a dense discretization of at least ten, but usually twenty-five points per wavelength is necessary for the extraction of a convergent solution. Therefore, the FDTD treatment of either electrically large geometries or fine detail structures typically results in a computationally intensive, memory and execution time consuming calculation. This fact naturally poses the significant question of deriving new, reasonably accurate techniques, with computational requirements that can be met by nowadays computer performance.

As an alternative to the conventional FDTD, several higher order numerical techniques have been developed [9], aimed at the discretization of electromagnetic structures at rates that may even approach the Nyquist limit. Furthermore, the incorporation of subgridding algorithms in the FDTD scheme [10] has indicated the potential of disconnecting the mean cell size within a domain, from the size of the smallest geometric detail that is contained in the latter. Yet, standard subgridding techniques involve spatiotemporal interpolations or extrapolations at the boundaries of different resolution parts of the numerical grid, that render the rigorous enforcement of the divergence free nature of the magnetic field and the continuity conditions a rather subtle issue.

If allowing for a relatively coarse mesh and taking into account localized geometric details are two challenges that novel numerical schemes are expected to meet, a third, but equally important one, is adaptivity. The concept of adaptivity implies the possibility of dynamic mesh refinement at regions of the computational domain that are electromagnetically active.

As the time-domain characterization of planar structures typically registers the history of a wideband (narrow) pulse propagation, adaptively imposing dense gridding conditions only in and around the pulse and the products of its retro-reflections can further extend the efficiency of a technique, beyond the limits that a static subgridding scheme can provide for.

Recently, wavelet-based time-domain methods, henceforth referred to as multiresolution time-domain (MRTD), employing Battle–Lemarie, Daubechies, biorthogonal, and Haar wavelets have been presented, for example, in [11–16] and references therein. According to MRTD, the electromagnetic field components are spatially expanded in a basis composed of scaling and wavelet functions, that effectively define a discrete mesh in space. In addition, pulse functions are used for the temporal field expansion. Upon substitution of these expansions into Maxwell's equations, Galerkin method is applied to derive finite difference equations with respect to the field scaling and wavelet spatiotemporal coefficients. These finite difference equations are solved by "marching in time," in the FDTD sense. It is noted that spatial variations of field components are reflected on the magnitude of their respective wavelet coefficients, which are either stored in computer memory or removed according to certain criteria (thresholding). In fact, several types of MRTD schemes with distinct numerical properties can be derived, depending on the wavelet basis that is employed for the discretization of Maxwell's equations.

In [11], Battle–Lemarie scaling functions enhanced by one wavelet level (zero-order wavelet functions) were employed for the formulation of the so-called W-MRTD scheme. Significant memory economy was shown to be associated to this method, because of its highly linear dispersion properties, that allowed for the modeling of electromagnetic structures at discretization rates that approached the Nyquist limit ($\lambda/2$). It is worth noting though, that these computational gains were strictly associated to the Battle–Lemarie scaling function rather than the Battle–Lemarie wavelet function basis. Furthermore, space adaptivity via an *a posteriori* thresholding of wavelet coefficients was pursued in [17]. Additionally, the range of W-MRTD applications that have been demonstrated so far, includes printed transmission line analysis [18], nonlinear circuit modeling [19], complex air-dielectric boundary problems, and antenna geometries [20].

Because of their relative simplicity, Haar MRTD schemes present an attractive alternative to their entire domain counterparts and have therefore become the subject of several studies [14, 21]. Since Haar MRTD constitutes the "wavelet extension" of the FDTD technique, a great deal of FDTD algorithms can be modified in order to resolve Haar MRTD modeling issues. Also, finite difference equations in Haar MRTD involve only "nearest neighbor" interactions, which is an attractive feature of an algorithm especially when it is combined with parallelization/domain decomposition techniques. For these reasons, the popularity of Haar MRTD has recently grown in the computational electromagnetics community.

It is worth noting that wavelet methods meet all aforementioned challenges for time-domain numerical schemes. First, by employing a high-order basis (such as cubic splines), a coarse mesh can be established, even when electrically small geometric features are to be included. Then, by adding wavelets and using adaptivity algorithms, similar to the ones that have been demonstrated in signal processing applications [22], a straightforward implementation of adaptive gridding is obtained. Moreover, an inherent contradiction related to MRTD schemes is that although their efficiency is expected to increase with the order of the multiresolution expansion at hand for homogeneous domains, the complexity of conductor, dielectric, and boundary modeling that they present, also increasing with wavelet order, seriously compromises their potential applicability to state-of-the-art devices. This complexity is associated with the incorporation of high-order wavelets into the finite difference expression of the constitutive relations for dielectrically inhomogeneous or conducting media.

This monograph is aimed at presenting two efficient solutions to the problem of establishing mesh adaptive time-domain numerical electromagnetic techniques.

First, a numerical interface between FDTD and the Haar wavelet-based MRTD is presented in detail, in Chapter 2. It is shown that the interface can greatly facilitate the application of both localized hard boundary conditions and the implementation perfectly matched layer (PML) absorbers for mesh truncation. It is worth noting that no interpolations or extrapolations (spatial or temporal) are necessary for the establishment of this interface. In addition, no stability issues are associated with its application.

Then, Chapter 3 deals with the most fundamental, application-oriented question, related to wavelet methods: *How* to efficiently implement the widely advertised mesh adaptivity, without defeating the purpose of obtaining operation savings. To this end, wavefront-tracking ideas are introduced and applied to one-dimensional linear and nonlinear optical wave propagation problems.

Following up to the technique of Chapter 3, Chapter 4 presents a computationally flexible and efficient method that can achieve mesh adaptivity, while fully preserving the versatility of FDTD. This technique, called the dynamically adaptive mesh refinement (AMR)-FDTD, essentially translates the AMR technique of computational fluid dynamics to three-dimensional electromagnetic problems. Its exceptional performance is demonstrated through microwave and optical applications in Chapters 5, 6.

CHAPTER 2

A Numerical Interface Between FDTD and Haar MRTD: Formulation and Applications

A numerically stable interface between the Haar wavelet-based MRTD technique and FDTD is presented in this section. Such a hybridization of MRTD facilitates the application of the method to open structures and inhomogeneous circuit geometries, where the use of high-order wavelets significantly complicates the formulation of a pure MRTD scheme. Furthermore, it allows for the straightforward enforcement of localized and perfectly matched layer (PML)-type absorbing boundary conditions. The fact that the implementation of the proposed interface involves no spatial or temporal interpolations indicates its potential to efficiently connect FDTD and Haar MRTD.

2.1 INTRODUCTION

The time-domain characterization of microwave structures, often encountered in wireless front-end applications such as filter, resonator or feed components, usually includes the modeling of fine detail complex boundaries and regions of dynamically varying field distributions. For this purpose, dense gridding conditions are necessary for the extraction of an accurate solution to the problem at hand. Nevertheless, the implementation of a uniformly dense mesh for an electrically large structure, translates to computationally burdensome simulations, as a result of both the size of the domain (in cells) and the stability-related restriction on the time step of a finite-difference time-domain (FDTD)-type of scheme, imposed by the small dimensions of the unit cell. Hence, the incorporation of local mesh refinement techniques into conventional time-domain solvers is motivated, as a means of alleviating the overall cost of modeling structures with localized geometric details.

Typically, subgridding techniques are implemented via spatiotemporal interpolations or extrapolations. Furthermore, a higher-order static subgridded FDTD algorithm, that needs no interpolatory operations has been recently proposed in [23]. However, wavelet-based numerical

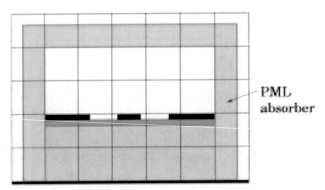

FIGURE 2.1: Coarse rectangular mesh for a layered coplanar waveguide structure, with cells containing variable dielectric permittivity and electric conductivity profiles

algorithms provide a natural framework for the implementation of dynamically adaptive subgridding, which holds the promise of significantly accelerating conventional CAD tools, even those with local subgridding features.

For homogeneous domains, the computational efficiency of wavelet-based schemes, such as MRTD, is expected to increase with the order of the multiresolution expansion [24]. Yet, the complexity of conductor, dielectric, and boundary modeling that they present, also increasing with wavelet order, seriously compromises their potential applicability to state-of-the-art devices. In particular, whenever a single cell includes variable material properties (such as dielectric and/or conducting layers), direct MRTD update equations are replaced by matrix expressions, resulting from the discretization of constitutive relations [11, 14]. Such cases are encountered in typical microwave devices, such as microstrip lines and coplanar waveguides (Fig. 2.1). In addition, the most effective mesh truncation technique nowadays, the perfectly matched layer (PML) absorber, is itself an inhomogeneous, uniaxially anisotropic (both electrically and magnetically) material.

The aforementioned contradiction is addressed here by means of a hybrid approach that connects the FDTD technique with the Haar wavelet-based MRTD, via a numerical interface. The purpose of this approach is to establish an efficient algorithm, allowing for the combination of the versatility of FDTD with the adaptivity of MRTD, employing the first in geometrically complex parts of the domain and the second in homogeneous regions.

In the past, interfaces between different numerical methods have been developed in order to address specific problems of interest. In [25], the frequency-domain method of moments (MoM) was coupled to FDTD, for the analysis of ground penetrating radar (GPR) problems, where GPR antenna operation was modeled by MoM, while the inhomogeneous ground (including dielectric stratification) was incorporated in an FDTD mesh. In [26], a Daubechies scaling function-based Wavelet–Galerkin scheme was coupled to FDTD for the modeling of wedge-loaded two-dimensional waveguide structures, apparently leading to significant computational savings compared to its pure FDTD counterpart. Still, the general applicability and efficiency of the scheme was questionable, given the largely different dispersion characteristics of the two methods. Moreover, a hybridization of the transmission line matrix (TLM) method and MRTD was pursued in [27], where the issue of dealing with disparate dispersion methods was explicitly addressed via a conditionally stable space–time interpolation scheme. Finally, [28] demonstrated a technique for including FDTD-modeled lumped elements in a three-dimensional domain simulated by the Haar wavelet-based MRTD formulation that was introduced in [14] and was restricted to one wavelet level.

This chapter proposes an interface between a Haar wavelet MRTD scheme of arbitrary number of wavelet levels and FDTD [29]. Under certain conditions, the two schemes are characterized by the same dispersion properties, a fact that is utilized in order to couple them with no interpolations or extrapolations and with absolutely no spurious reflections at their interface.

The structure of this chapter is as follows: First, a brief overview of multiresolution analysis (that the MRTD technique is based upon) is provided, along with a presentation of how the technique can be systematically formulated. Then, the formulation of a Haar MRTD scheme (supporting an arbitrarily high spatial resolution) is presented. Noting that the dispersion equation for the latter coincides with the dispersion equation of an FDTD scheme of the same spatial resolution, a simple connection algorithm between the two is proposed. Emphasis is placed on the implementation problems that arise in the application of update equations at the boundaries of the two domains. Validation studies include the analysis of two-dimensional dielectric cavity structures, under various FDTD/MRTD mesh configurations. Finally, the method is applied for mesh truncation in open domain problems and the modeling of structures with metal inserts such as a fin-loaded cavity.

2.2   MULTIRESOLUTION ANALYSIS: A BRIEF OVERVIEW

The theoretical foundation of wavelet-based numerical techniques is the so-called multiresolution analysis theory, that is often encountered under the headline of wavelets. A brief overview of this subject is provided here for the purpose of completeness. A detailed study of the topic is contained in [30].

2.2.1   General

Wavelet bases are a mathematical tool for hierarchical decomposition of functions in an orthogonal expansion, according to the general scheme:

$$f(\xi) = \sum_k c_k \, \phi(\xi - k) + \sum_j \sum_k d_{j,k} \, \psi(2^j \, \xi - k) \qquad (2.1)$$

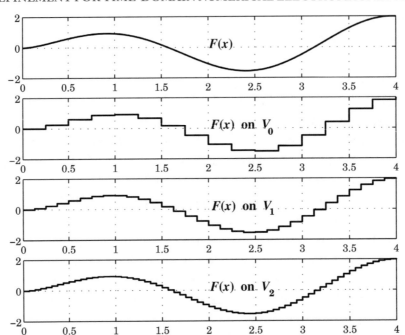

FIGURE 2.2: Projections of a function f on three successive approximation spaces V_0, V_1, V_2

where f is assumed to belong to $\mathcal{L}^2(\mathbf{R})$.[1] In (2.1), the first sum represents the projection of $f(\xi)$ onto a subspace V_0, that corresponds to an approximation of f at a "coarse" level of resolution. The basis of V_0 is generated by orthogonal translations of $\phi(\xi)$ which is called the *scaling* function. The resolution of V_0 is successively refined by the second sum, which consists of projections of f onto the subspaces W_j, each one being spanned by a *wavelet* basis $\{\psi(2^j\xi - k)\}$. The function $\psi(\xi)$ is called the *mother wavelet*, since all other wavelet basis functions $\psi_{j,k}$ are simply produced by dilations (for adjustment of resolution) and translations of ψ. It is noted that a basic property of the subspaces V_0, W_0 is

$$V_0 \bigoplus W_0 = V_1 \qquad (2.2)$$

where V_1 corresponds to a level of resolution twice that of V_0. This means that scaling and zero wavelet functions form a basis of orthogonal functions that spans V_1. The notion of a projection of f onto approximation spaces at successive levels of resolution is explained in Fig. 2.2.

Recursively,

$$V_0 \bigoplus W_0 \bigoplus W_1 \cdots \bigoplus W_{k-1} = V_k \qquad (2.3)$$

---

[1]Hilbert space of square integrable functions.

In other words, adding one wavelet level, is equivalent to improving the resolution of an approximation, like the one in Eq. (2.1), by a factor of two.[2] Hence, any desired resolution of approximation can be achieved by adding an appropriate number of wavelet levels.

The advantage of this approach lies in the dependence of the magnitude of wavelet coefficients on the local regularity of f. In particular, wavelet coefficients assume large values near discontinuities or singularities. Indeed, an explicit relationship between the magnitude and the "Lipschitz uniformity" of a square integrable function [31] is given by the following theorem [32]: *Let $0 < \alpha < n$ be a real number that is not an integer. The function $f(x)$ is uniformly Lipschitz of order α over the interval $[a, b]$ if and only if for any $n \in \mathbf{Z}$ and $j \in \mathbf{Z}$ such that $2^{-j} n \in (a, b), |< f, \psi_{j,n} >| = \mathcal{O}\left(2^{-(\alpha+1/2)j}\right)$.*[3]

Therefore, once a numerical analysis of a problem is performed by expanding the unknown function in a wavelet basis, high-order wavelet coefficients, which are located away from discontinuities or singularities, typically assume insignificant values, that can be *thresholded* to zero. *Thresholding* of a wavelet coefficient implies its elimination from subsequent operations, or—equivalently—it's being treated as if it were equal to zero. By this procedure, an adaptive, temporally moving mesh is implemented.

2.2.2   Wavelet Bases

The simplest (and oldest) orthogonal wavelet system is the *Haar* basis [33]. Its study is useful from a theoretical point of view because it offers an intuitive understanding of many multiresolution properties. Furthermore, due to its simplicity, this basis is widely employed in a series of applications, hence its study is also practically interesting.

The Haar scaling function is defined as a pulse of unit length:

$$\phi(\xi) = \begin{cases} 1, & 0 \le \xi < 1 \\ 0, & \text{otherwise} \end{cases} \tag{2.4}$$

and the scaling function basis is produced by orthogonal translations of the latter:

$$\phi_k(\xi) = \phi(\xi - k) \tag{2.5}$$

The Haar mother wavelet function is defined as:

$$\psi(\xi) = \begin{cases} 1, & 0 \le \xi < 1/2 \\ -1, & 1/2 \le \xi < 1 \\ 0, & \text{otherwise.} \end{cases} \tag{2.6}$$

---

[2]Dyadic multiresolution analyses are considered here. In general, modifying the dilation factor for the generation of the wavelet basis from the mother wavelet can yield other types of MRA.

[3]The notation $< f, g >$ implies inner product. If $f(h)$ and $g(h)$ are two functions of h, it is said that $f(h) = \mathcal{O}(g(h))$, as $h \to 0$, if there is some constant C, such that: $|f(h)| < C|g(h)|$, for h sufficiently small.

The wavelet basis is then produced by both translations *and* dilations, as shown in the definition:

$$\psi_{j,k}(\xi) = 2^{j/2}\psi(2^j\xi - k). \tag{2.7}$$

It is noted that the same definition holds for the derivation of any dyadic wavelet basis, from its generating mother wavelet. In all these definitions ξ is a normalized, dimensionless variable. For example, if ϕ or ψ functions represent a variation along the x-direction of a computational mesh with a cell size Δx, then simply: $\xi = x/\Delta x$. The Haar scaling and mother wavelet functions are depicted in Fig. 2.3, and the mechanism, in which multiresolution analysis is brought about by employing this basis (in the sense of Eq. (2.3)), is qualitatively explained in Fig. 2.4.

The Haar basis has an excellent localization in space (or time) domain and poor localization in the corresponding Fourier domain, as shown in Fig. 2.5. Therefore, the Haar basis does not share the typical wavelet property of combining good localization in both domains (within the limits of the uncertainty principle [22]).

A basis that has also been used in applications, is the cubic spline Battle–Lemarie basis [34]. The Battle–Lemarie scaling and mother wavelet (Fig. 2.6) are entire domain functions and therefore schemes that are developed in this basis have to be truncated with respect to space. However, Battle–Lemarie scaling and wavelets have an excellent localization both in space and Fourier domains (Fig. 2.7), a feature that permits an *a priori* estimate of the necessary levels of resolution for correct field modeling.

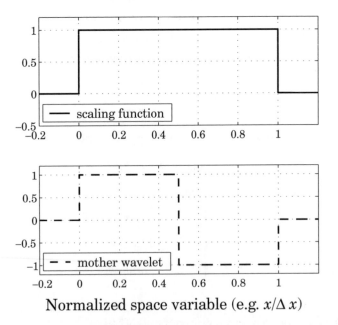

FIGURE 2.3: Haar scaling and mother wavelet function

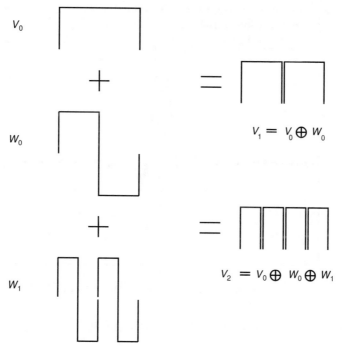

FIGURE 2.4: Demonstration of the multiresolution principle as implemented by the Haar basis

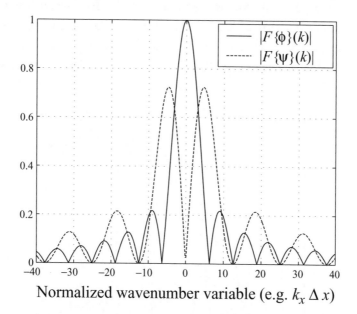

Normalized wavenumber variable (e.g. $k_x \Delta x$)

FIGURE 2.5: Haar scaling and mother wavelet functions in the Fourier domain

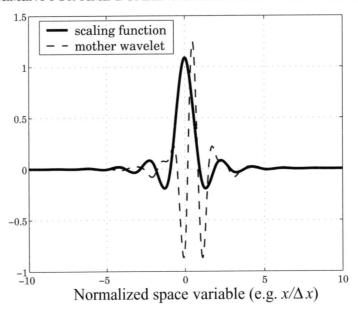

FIGURE 2.6: Battle–Lemarie scaling and mother wavelet function

Figures 2.5 and 2.7 also provide an intuitive explanation of the successive approximation procedure that is connected to multiresolution analysis. Evidently, the scaling function has the Fourier domain pattern of a *low—pass filter*, while the mother wavelet is by the same token a *band—pass filter*. Hence, scaling functions alone provide a description of "low frequency" characteristics of a signal, while the addition of wavelets enhances the capability of

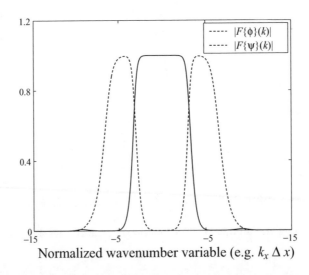

FIGURE 2.7: Battle–Lemarie scaling and mother wavelet functions in the Fourier domain

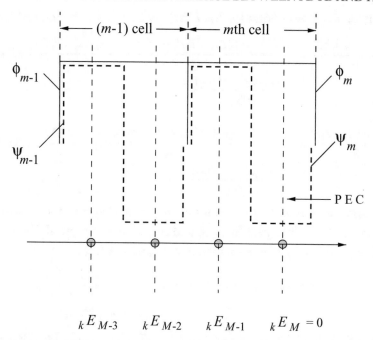

FIGURE 2.8: Equivalent grid points (e.g.p.) within a scaling cell as introduced by a first-order wavelet expansion in x-, z-directions

a scheme to describe the high-frequency content of a signal. This argument also explains the adaptivity property of wavelet-based numerical schemes: wavelet coefficients correspond to high-frequency signal variations, in the absence of which, the value of these coefficients decays to insignificant levels.

2.3   DERIVATION OF TIME-DOMAIN SCHEMES BY THE METHOD OF MOMENTS

In [35], a systematic way to formulate time-domain numerical schemes for Maxwell's equations is provided. This technique, which is also employed in the present monograph for the derivation of MRTD schemes, is explained here, by considering first, the following one-dimensional wave equation example:

$$\frac{\partial E(z, t)}{\partial z} = \frac{1}{c} \frac{\partial E(z, t)}{\partial t} \tag{2.8}$$

For the numerical solution of this equation, a discrete space–time mesh is introduced and the values of the electric field $E(m\Delta z, k\Delta t) = {_k}E_m$ are sought by marching in time. The FDTD scheme for this equation is derived by approximating the partial derivatives of each side of the equation at a space–time mesh point $(m \, \Delta z, k \, \Delta t)$, by centered differencing [36]:

$$\frac{\partial E(m \Delta z, k \Delta t)}{\partial z} = \frac{E((m+1)\Delta z, k\Delta t) - E((m-1)\Delta z, k\Delta t)}{2\Delta z} + O(\Delta z^2)$$

$$\approx \frac{{}_k E_{m+1} - {}_k E_{m-1}}{2\Delta z} \qquad (2.9)$$

$$\frac{\partial E(m \Delta z, k \Delta t)}{\partial t} = \frac{E(m\Delta z, (k+1)\Delta t) - E(m\Delta z, (k-1)\Delta t)}{2\Delta t} + O(\Delta t^2)$$

$$\approx \frac{{}_{k+1} E_m - {}_{k-1} E_m}{2\Delta t} \qquad (2.10)$$

The previous expressions stem from a reformulation of a Taylor series expansion for $E(z, t)$ around $m \, \Delta z$ and $k \, \Delta t$, with respect to space and time. Then, simple algebraic manipulations lead to the following second-order accurate approximation of (2.8):

$$_{k+1}E_m = {}_{k-1}E_m + \frac{c\,\Delta t}{\Delta z} \left({}_k E_{m+1} - {}_k E_{m-1} \right) \qquad (2.11)$$

$$= {}_{k-1}E_m + s \quad \left({}_k E_{m+1} - {}_k E_{m-1} \right) \qquad (2.12)$$

where $s = c \, \Delta t / \Delta z$ is the CFL (Courant–Friedrichs–Levy) number [36]. For this scheme to be stable, the CFL number has to be less than one. Alternatively, for the discretization of (2.8) via the method of moments, the electric field $E(z, t)$ is expanded in pulse basis functions (or equivalently: Haar scaling functions) $\{b_n(z)\}$, $\{b_n(t)\}$ in space and time:

$$E(z, t) = \sum_{k, m} {}_k E_m \, b_m(z) \, b_k(t) \qquad (2.13)$$

with

$$b_n(x) = \begin{cases} 1, & n\Delta x < x < (n+1)\Delta x \\ 0, & \text{otherwise} \end{cases} \qquad (2.14)$$

Substituting into (2.8), yields:

$$\sum_{k, m} {}_k E_m \frac{d b_m(z)}{dz} \, b_k(t) = \frac{1}{c} \sum_{k, m} {}_k E_m \, b_m(z) \frac{d b_k(t)}{dt} \qquad (2.15)$$

Then, (2.15) is sampled in space and time using the complex conjugate of the basis functions themselves as testing functions (Galerkin method). To carry out the testing procedure, the following integrals are employed:

$$\int_{-\infty}^{+\infty} b_n(x) \frac{d b_{n'}(x)}{dx} dx = \frac{1}{2} (\delta_{n',n+1} - \delta_{n',n-1}) \qquad (2.16)$$

$$\int_{-\infty}^{+\infty} b_n(x) \, b_{n'}(x) \, dx = \delta_{n',n} \, \Delta x \qquad (2.17)$$

where $\delta_{n',n}$ is Kronecker's delta:

$$\delta_{n',n} = \begin{cases} 1, & \text{if } n' = n \\ 0, & \text{if } n' \neq n \end{cases}$$

Hence, testing the left hand-side of (2.15), yields:

$$\int_{-\infty}^{+\infty}\int_{-\infty}^{+\infty} dz\, dt\, h_m(z)\, h_k(t)\, \frac{\partial E(z,t)}{\partial z} = \sum_{k',m'} {}_{k'}E_{m'}\, (\delta_{m',m+1} - \delta_{m',m-1})\, \delta_{k',k}\, \frac{\Delta t}{2}$$

$$= ({}_kE_{m+1} - {}_kE_{m-1})\frac{\Delta t}{2}. \tag{2.18}$$

Similarly, testing the right hand-side of (2.15):

$$\int_{-\infty}^{+\infty}\int_{-\infty}^{+\infty} dz\, dt\, h_m(z)\, h_k(t)\, \frac{1}{c}\frac{\partial E(z,t)}{\partial t} = \sum_{k',m'} {}_{k'}E_{m'}\, \delta_{m',m}\, (\delta_{k',k+1} - \delta_{k',k-1})\, \frac{\Delta z}{2c}$$

$$= ({}_{k+1}E_m - {}_{k-1}E_m)\frac{\Delta z}{2c}. \tag{2.19}$$

Thus, it is easily concluded that Eq. (2.12) is again derived and therefore the two methods for approximating (2.8) in a discrete space, are shown to be equivalent.

In a similar manner, expanding the electric and magnetic fields in scaling and wavelet functions and applying the method of moments for the discretization of Maxwell's equations (as indicated in the previous section), the MRTD scheme is derived. In this case, the field expansion in (2.13) is written as:

$$E(z,t) = \sum_{k,\,m} {}_kE_m^{\phi}\, \phi_m(z)\, h_k(t) + \sum_{r=0}^{R_{\max}}\sum_{p=0}^{2^r-1} {}_kE_m^{\psi_{r,p}}\, \psi_{m,\,p}^r(z)\, h_k(t), \tag{2.20}$$

where $R_{\max}$ is the maximum wavelet-order that has been introduced. Also,

$$\phi_m(z) = \phi\left(\frac{z}{\Delta z} - m\right) \tag{2.21}$$

is a scaling basis produced by a scaling function ϕ and:

$$\psi_{m,\,p}^r(z) = 2^{r/2}\, \psi\left(2^r\left(\frac{z}{\Delta z} - m\right) - p\right) \tag{2.22}$$

with $p = 0, 1, \ldots, 2^r - 1$. Hence, a slightly modified definition with respect to (2.7) is used for the wavelet basis here, in order to keep a correspondence between the cell index m of the scaling and the wavelet functions. This definition is schematically explained for the case of the Haar basis in Fig. 2.9.

The introduction of wavelets gives rise to a refinement of the scaling function-based approximation of $E(z,t)$ by a factor $2^{R_{\max}+1}$. To explain the mechanics of this mesh refinement,

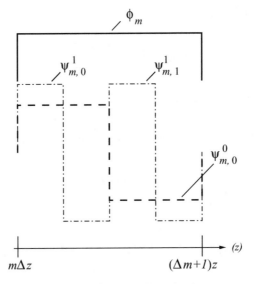

FIGURE 2.9: Explanation of definition (2.22) for the Haar basis

the simple example of the Haar basis is employed. In Fig. 2.10, two scaling cells, defined by pulse functions are shown. Assuming the use of a first-order scheme (two wavelet levels), the following wavelet terms are included within the nth cell: $\psi_{n,0}^0$, $\psi_{n,0}^1$, $\psi_{n,1}^1$. Linear combinations of their coefficients with the scaling coefficient, produce four *equivalent grid points* (e.g.p.) within the cell. The electric field values at the equivalent grid points and the positions of the

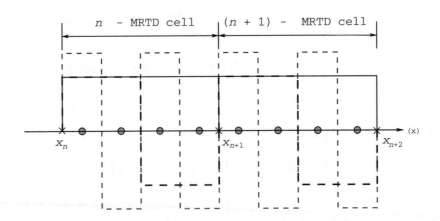

FIGURE 2.10: Mesh refinement in the Haar basis

TABLE 2.1: Position of and Field Values at the MRTD Equivalent Grid Points of Fig. 2.10

POSITION	FIELD NODAL VALUE
$x_n + 0.125\,\Delta x$	$E_n^\phi + E_n^{\psi_{0,0}} + \sqrt{2}E_n^{\psi_{1,0}}$
$x_n + 0.375\,\Delta x$	$E_n^\phi + E_n^{\psi_{0,0}} - \sqrt{2}E_n^{\psi_{1,0}}$
$x_n + 0.625\,\Delta x$	$E_n^\phi - E_n^{\psi_{0,0}} + \sqrt{2}E_n^{\psi_{1,1}}$
$x_n + 0.875\,\Delta x$	$E_n^\phi - E_n^{\psi_{0,0}} - \sqrt{2}E_n^{\psi_{1,1}}$

latter within the nth cell are provided in Table 2.1. A mathematically elegant and compact way to express the relationship between field nodal values and MRA expansion coefficients is the following matrix formulation:

$$
\begin{bmatrix} \hat{E}_n^1 \\ \hat{E}_n^2 \\ \hat{E}_n^3 \\ \hat{E}_n^4 \end{bmatrix} = \begin{bmatrix} +1 & +1 & +\sqrt{2} & 0 \\ +1 & +1 & -\sqrt{2} & 0 \\ +1 & -1 & 0 & +\sqrt{2} \\ +1 & -1 & 0 & -\sqrt{2} \end{bmatrix} \cdot \begin{bmatrix} E_n^\phi \\ E_n^{\psi_{0,0}} \\ E_n^{\psi_{1,0}} \\ E_n^{\psi_{1,1}} \end{bmatrix}
\tag{2.23}
$$

The matrix on the right hand-side of (2.23) is the one-dimensional wavelet transform matrix for a two-level wavelet scheme. This formulation will be later on employed in the explicit enforcement of localized boundary conditions and the development of a numerical interface between the FDTD method and MRTD.

The modified finite difference equations for the MRTD approximation of (2.8) are now dependent on the form that the following integrals assume in a given wavelet basis:

$$
I_{n,n'}^{\phi,\phi} = \int_{-\infty}^{+\infty} \phi_n(z)\,\frac{d\phi_{n'}(z)}{dz}\,dz
$$

$$
I_{n,n'}^{\phi,\psi_p^r} = \int_{-\infty}^{+\infty} \phi_n(z)\,\frac{d\psi_{n',p}^r(z)}{dz}\,dz
$$

$$
I_{n,n'}^{\psi_p^r,\phi} = \int_{-\infty}^{+\infty} \psi_{n,p}^r(z)\,\frac{d\phi_{n'}(z)}{dz}\,dz
\tag{2.24}
$$

$$
I_{n,n'}^{\psi_p^r,\psi_p^r} = \int_{-\infty}^{+\infty} \psi_{n,p}^r(z)\,\frac{d\psi_{n',p'}^{r'}(z)}{dz}\,dz\,.
$$

Letting $l = m' - m$, the following set of update equations for scaling and wavelet terms, cast in a generic form, is deduced:

$$
\begin{aligned}
{}_{k+1}E_m^\phi &= {}_{k-1}E_m^\phi + \frac{2c\,\Delta t}{\Delta z} \sum_{l=-\infty}^{+\infty} \left\{ I_{m,m+l}^{\phi,\,\phi}\,{}_kE_{m+l}^\phi + \sum_{r=0}^{R_{\max}}\sum_{p=0}^{2^r-1} I_{m,m+l}^{\phi,\,\psi_p^r}\,{}_kE_{m+l}^{\psi_{r,p}} \right\} \\[2mm]
{}_{k+1}E_m^{\psi_{r,p}} &= {}_{k-1}E_m^{\psi_{r,p}} + \frac{2c\,\Delta t}{\Delta z} \sum_{l=-\infty}^{+\infty} \left\{ I_{m,m+l}^{\psi_{r,p},\,\phi}\,{}_kE_{m+l}^\phi + \sum_{r'=0}^{R_{\max}}\sum_{p'=0}^{2^r-1} I_{m,m+l}^{\psi_p^r,\,\psi_{p'}^{r'}}\,{}_kE_{m+l}^{\psi_{r',p'}} \right\}
\end{aligned}
\tag{2.25}
$$

The infinite sums with respect to the position index in (2.25), are either reduced to finite ones, if finite domain basis functions are involved, or simply approximated as finite ones, if entire domain basis functions are involved. A term widely used in the wavelet literature, as a formal substitute for the concept of entire or finite domain functions is that of *support*. The definition of this term is provided here, for completeness: The support of a function f is the closure of the set $\{x : f(x) \neq 0\}$, or $\overline{\{x : f(x) \neq 0\}}$ [31]. In the case of infinitely supported basis functions (e.g., the Battle–Lemarie basis employed in [11]), the variation of the index l of the infinite sums is restricted within a finite range $[-N, N]$, where N is called the *stencil* of the method.

The computation of the integrals of (2.24) can be carried out either analytically or computationally, depending on the integrands themselves. It is often the case, that wavelet functions are defined in the Fourier domain, rather than directly. Therefore, Fourier calculus has to be employed for the extraction of their value.

As a final note, the clarification of the use of the term "order" of an MRTD scheme, as opposed to its common use in numerical methods, is necessary. The numerical discretization of (2.8), given in (2.12), is called *second-order* accurate, or—simply— a *second-order scheme*, because it implies a numerical approximation of the spatial and temporal derivatives of (2.8), with an error proportional to Δz^2 and Δt^2, respectively. This fact becomes evident by pure inspection of expressions (2.10). On the other hand, an MRTD scheme is of nth order, if wavelets up to order n are introduced. It is noted that wavelets are solely connected to mesh refinement and do not affect the order of accuracy of the underlying scheme.

2.4   TWO-DIMENSIONAL HYBRID ARBITRARY-ORDER HAAR MRTD/FDTD SCHEME: FORMULATION

2.4.1   MRTD, FDTD Equations

In this section, MRTD is applied to the following system of two-dimensional TE_z Maxwell's equations:

$$
\frac{\partial E_y}{\partial t}(\overline{\rho},\, t) = \frac{1}{\epsilon}\left(\frac{\partial H_x}{\partial z}(\overline{\rho},\, t) - \frac{\partial H_z}{\partial x}(\overline{\rho},\, t) \right)
\tag{2.26}
$$

$$\frac{\partial H_x}{\partial t}(\overline{\rho},\, t) = \frac{1}{\mu}\frac{\partial E_y}{\partial z}(\overline{\rho},\, t) \tag{2.27}$$

$$\frac{\partial H_z}{\partial t}(\overline{\rho},\, t) = -\frac{1}{\mu}\frac{\partial E_y}{\partial x}(\overline{\rho},\, t) \tag{2.28}$$

with $\overline{\rho} = x\hat{x} + z\hat{z}$. Based on the method outlined in [11], update equations are derived by the method of moments, assuming a spatial expansion of electromagnetic field components in scaling and wavelet functions of an arbitrary basis and up to arbitrary orders $r_{x,max}$, $r_{z,max}$ in x-, z-directions respectively. However, our analysis is restricted to dyadic wavelet transforms, as they are the most commonly used for the purpose of adaptively solving partial differential equations. In the subsequent development, the discretization of a two-dimensional domain (in which field solutions are sought) in cells of Δx by Δz is pursued, by means of a wavelet basis, defined by the scaling function ϕ and the so-called mother wavelet ψ [30]. Then, $\phi_m(\xi) = \phi(\xi/\Delta\xi - m)$ denotes the mth scaling function in $\xi = x$-, z-directions. Accordingly, the wavelet functions of order r that recursively refine the resolution of ϕ_m are defined as: $\psi_{m,p}^r = 2^{r/2}\psi(2^r(x/\Delta x - m) - p)$, where $p = 0, \cdots 2^r - 1$. A basis of pulse functions $h_n(t) = h(t/\Delta t - n)$ (defined as in [11]), is employed for field expansion in time, where Δt denotes the time step, limited by the choice of Δx and Δz through the stability condition. Given these definitions, E_y, H_x, H_z are expressed in the form of the following orthogonal expansions:

$$
\begin{aligned}
E_y(\overline{\rho},\, t) = \sum_n h_n(t) \sum_{i,m} \Big\{ &\,_nE_{i,m}^{y,\,\phi\phi}\phi_i(x)\,\phi_m(z) \\
&+ \sum_{r_z,p_z} {}_nE_{i,m}^{y,\,\phi\,\psi_{r_z,p_z}}\phi_i(x)\,\psi_{m,p_z}^{r_z}(z) \\
&+ \sum_{r_x,p_x} {}_nE_{i,m}^{y,\,\psi_{r_x,p_x}\phi}\psi_{i,p_x}^{r_x}(x)\,\phi_m(z) \\
&+ \sum_{r_x,p_x}\sum_{r_z,p_z} {}_nE_{i,m}^{y,\,\psi_{r_x,p_x}\psi_{r_z,p_z}}\psi_{i,p_x}^{r_x}(x)\,\psi_{m,p_z}^{r_z}(z) \Big\}
\end{aligned}
\tag{2.29}
$$

$$
\begin{aligned}
H_x(\overline{\rho},\, t) = \sum_n h_{n'}(t) \sum_{i,m} \Big\{ &\,_{n'}H_{i,m'}^{x,\,\phi\phi}\phi_i(x)\,\phi_{m'}(z) \\
&+ \sum_{r_z,p_z} {}_{n'}H_{i,m'}^{x,\,\phi\,\psi_{r_z,p_z}}\phi_i(x)\,\psi_{m',p_z}^{r_z}(z) \\
&+ \sum_{r_x,p_x} {}_{n'}H_{i,m'}^{x,\,\psi_{r_x,p_x}\phi}\psi_{i,p_x}^{r_x}(x)\,\phi_{m'}(z) \\
&+ \sum_{r_x,p_x}\sum_{r_z,p_z} {}_{n'}H_{i,m'}^{x,\,\psi_{r_x,p_x}\psi_{r_z,p_z}}\psi_{i,p_x}^{r_x}(x)\,\psi_{m',p_z}^{r_z}(z) \Big\}
\end{aligned}
\tag{2.30}
$$

$$H_z(\bar{\rho}, t) = \sum_n h_{n'}(t) \sum_{i,m} \left\{ {}_{n'} H_{i',m}^{z,\,\phi\,\phi} \phi_{i'}(x)\, \phi_m(z) \right.$$

$$+ \sum_{r_z, p_z} {}_{n'} H_{i',m}^{z,\,\phi\,\psi_{r_z, p_z}} \phi_{i'}(x)\, \psi_{m, p_z}^{r_z}(z)$$

$$+ \sum_{r_x, p_x} {}_{n'} H_{i',m}^{z,\,\psi_{r_x, p_x}\,\phi} \psi_{i', p_x}^{r_x}(x)\, \phi_m(z)$$

$$\left. + \sum_{r_x, p_x} \sum_{r_z, p_z} {}_{n'} H_{i',m}^{z,\,\psi_{r_x, p_x}\,\psi_{r_z, p_z}} \psi_{i', p_x}^{r_x}(x)\, \psi_{m, p_z}^{r_z}(z) \right\} \qquad (2.31)$$

where $n' = n + 1/2$, $i' = i + s_x$, $m' = m + s_z$. Thus, while half a time step offset between the update of electric and magnetic field terms is kept (as in FDTD), the offset of electric and magnetic scaling cells in x- and z-directions is left as a parameter under investigation. Most MRTD studies, with the notable exception of [37], adopt the choice of $s_x = s_z = 1/2$, based on the FDTD practice. In this section, a systematic way of determining these offsets is set forth, by introducing the notion of *equivalent MRTD grid points*.

Assume that a certain scaling function basis generates electric field grid points in one dimension ($\xi = x,\ z$) : $i \cdot \Delta\xi$, $i = 1, 2, \ldots, N_\xi$. Then, the introduction of wavelets of orders $r = 0, 1, \ldots, r_{\xi,\max}$, refines the mesh in ξ-direction by a factor of $\rho_\xi = 2^{r_{\xi,\max}+1}$, since each wavelet level successively doubles the resolution of the underlying approximation. In mathematical terms, the E_y-expansion of (2.30) can be cast in the next equivalent form [30]:

$$E_y(\bar{\rho}, t) = \sum_n h_n(t) \sum_{i,m} \left\{ {}_n E_{i,m}^{y,\,\phi^{R_x}\,\phi^{R_z}} \phi_i^{R_x}(x)\, \phi_m^{R_z}(z) \right\}, \qquad (2.32)$$

where $R_x = r_{x,\max} + 1$, $R_z = r_{z,\max} + 1$ and $\{\phi_n^R(\xi)\} = \{2^{R/2}\phi(2^R(\xi/\Delta\xi) - n)\}$ is the scaling basis that produces by itself an approximation of the same resolution as (2.30). In fact, the latter corresponds to the hierarchical multiresolution decomposition of the former and the coefficients in (2.30) can directly be deduced from the ones in (2.32) via the wavelet transform. This wavelet-induced mesh refinement can also be perceived as a procedure of generating equivalent grid points that give rise to a mesh of cell sizes $\Delta x/\rho_x$, $\Delta z/\rho_z$. Similar observations hold for (2.31), (2.31).

This argument is demonstrated for a zero-order Haar MRTD scheme, utilizing Haar scaling and zero-order wavelet functions (Fig. 2.3), in Fig. 2.11 and 2.12, which depict the equivalent grid points that are generated in both cases, in one dimension. It is noted that in general, linear combinations of scaling and wavelet functions yield electric field values at points: $(i + (p + 0.5)/2^{r_{x,\max}+1})\,\Delta x$, with $p = 0, 1 \cdots 2^{r_{x,\max}+1} - 1$. In our two-dimensional example, these figures represent z-*cuts* of the mesh, including E_y and H_z grid points (which are necessary

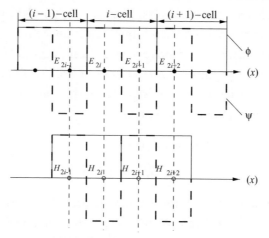

FIGURE 2.11: Electric/magnetic field equivalent grid points for zero-order Haar MRTD in one dimension, under the common convention of half a cell offset between electric/magnetic scaling cells

for the approximation of x-partial derivatives involved in E_y-updates). Up to a normalization multiplicative constant, the field values at equivalent grid points within each cell, are computed as the sum and the difference of scaling and zero-order wavelet terms, respectively. Accordingly, H_z equivalent grid points are located at $\left(i + s_x + (p + 0.5)/2^{r_{x,\max}+1}\right)\Delta x$. However, the purpose

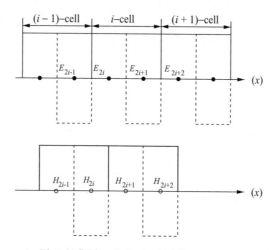

FIGURE 2.12: Electric/magnetic field equivalent grid points for zero-order Haar MRTD in one dimension, with electric/magnetic scaling cell offset chosen in consistence with (2.33)

of using wavelets is to implement in this "new" mesh of equivalent grid points, the method produced by scaling functions only, at a resolution that is finer (in x-direction) by the wavelet refinement factor ρ_x. Hence, if this method defines half a scaling cell offset between electric and magnetic nodes, the wavelet augmented method has to define half an *equivalent* cell offset between the equivalent electric/magnetic grid points in this direction. The choice of s_x and s_z that is consistent with this requirement is given below:

$$s_x = 0.5\rho_x = 2^{-r_{x,\max}-2}$$
$$s_z = 0.5\rho_z = 2^{-r_{z,\max}-2}$$
(2.33)

Figure 2.11 shows the node arrangement in zero-order Haar MRTD, under the common convention of half a cell offset between electric/magnetic scaling cells. Apparently, equivalent electric and magnetic equivalent nodes are now collocated. On the other hand, if the separation of electric and magnetic scaling cells is chosen in consistence with (2.33), which in this case yields an offset of one quarter of a cell (Fig. 2.12), the equivalent grid points are leap-frogged in space and correctly correspond to the mesh of an FDTD scheme of cell size $\Delta x/2$. As discussed in [38], this latter approach leads to MRTD schemes with consistent numerical dispersion properties (as opposed to the former approach) and therefore, it is adopted in the following derivations.

Upon substitution of the MRTD expansions of all field components into Eq. (2.26)–(2.28), Galerkin's method is applied for the derivation of field update equations. In the following, the update equations for E_y, $b_x = H_x \Delta x$ and $b_z = H_z \Delta z$ coefficients are provided. The scaling cell dimensions are $\Delta x \times \Delta z$. Also, $n' = n + 1/2$, $i' = i + s_x$, $m' = m + s_z$, $\zeta = \psi_{r_x,p_x}$ and $\eta = \psi_{r_z,p_z}$. Finally, the coefficients $\mathcal{D}_0$, $\mathcal{D}_1$ are given in Section 2.8.

E_y-**update equations**

$$
\begin{aligned}
_{n+1}E_{i,m}^{y,\phi\phi} = {}_{n}E_{i,m}^{y,\phi\phi} + \frac{\Delta t}{\epsilon \Delta x \Delta z} & \left\{ {}_{n'}b_{i,m'}^{x,\phi\phi} - {}_{n'}b_{i,m'-1}^{x,\phi\phi} + \sum_r 2^{r/2}\left({}_{n'}b_{i,m'-1}^{x,\phi\psi_{r,2^r-1}} - {}_{n'}b_{i,m'}^{x,\phi\psi_{r,2^r-1}} \right) \right. \\
& \left. - \left[{}_{n'}b_{i',m}^{z,\phi\phi} - {}_{n'}b_{i'-1,m}^{z,\phi\phi} + \sum_r 2^{r/2}\left({}_{n'}b_{i'-1,m}^{z,\psi_{r,p}\phi} - {}_{n'}b_{i',m}^{z,\psi_{r,p}\phi} \right) \right] \right\}
\end{aligned}
$$
(2.34)

$$
\begin{aligned}
_{n+1}E_{i,m}^{y,\phi\psi_{r,p}} = {}_{n}E_{i,m}^{y,\phi\psi_{r,p}} + \frac{\Delta t}{\epsilon \Delta x \Delta z} & \left\{ 2^{r/2}\delta_{p,0}\left({}_{n'}b_{i,m'}^{x,\phi\phi} - {}_{n'}b_{i,m'-1}^{x,\phi\phi} \right) \right. \\
& + \sum_{r',p'}\left[\mathcal{D}_2(r,p,r',p')\,{}_{n'}b_{i,m'}^{x,\phi\psi_{r',p'}} + \mathcal{D}_3(r,p,r',p')\,{}_{n'}b_{i,m'-1}^{x,\phi\psi_{r',p'}} \right] \\
& \left. - \left[{}_{n'}b_{i',m}^{z,\phi\psi_{r,p}} - {}_{n'}b_{i'-1,m}^{z,\phi\psi_{r,p}} + \sum_q 2^{q/2}\left({}_{n'}b_{i',m-1}^{z,\psi_{q,2q-1}\psi_{r,p}} - {}_{n'}b_{i'-1,m}^{z,\psi_{q,2q-1}\psi_{r,p}} \right) \right] \right\}
\end{aligned}
$$
(2.35)

$$
{}_{n+1}E_{i,m}^{y,\psi_{r,p}\phi} = {}_nE_{i,m}^{y,\psi_{r,p}\phi} + \frac{\Delta t}{\epsilon \Delta x \Delta z}\bigg\{ {}_{n'}h_{i,m'}^{x,\psi_{r,p}\phi} - {}_{n'}h_{i,m'-1}^{x,\psi_{r,p}\phi}
$$

$$
+ \sum_q 2^{q/2}\left({}_{n'}h_{i,m'-1}^{x,\psi_{r,p}\psi_{q,2^q-1}} - {}_{n'}h_{i,m'}^{x,\psi_{r,p}\psi_{q,2^q-1}}\right) - \bigg[2^{r/2}\delta_{p,0}\left({}_{n'}h_{i',m}^{z,\phi\phi} - {}_{n'}h_{i'-1,m}^{z,\phi\phi}\right)
$$

$$
+ \sum_{r',p'}\bigg[\mathcal{D}_2(r,p,r',p')\,{}_{n'}h_{i',m}^{z,\psi_{r',p'}\phi} + \mathcal{D}_3(r,p,r',p')\,{}_{n'}h_{i'-1,m}^{z,\psi_{r',p'}\phi}\bigg)\bigg]\bigg\} \qquad (2.36)
$$

$$
{}_{n+1}E_{i,m}^{y,\psi_{r,p}\psi_{q,w}} = {}_nE_{i,m}^{y,\psi_{r,p}\psi_{q,w}} + \frac{\Delta t}{\epsilon \Delta x \Delta z}\bigg\{ 2^{q/2}\delta_{w,0}\left({}_{n'}h_{i,m'}^{x,\psi_{r,p}\phi}\,{}_{n'}h_{i,m'-1}^{x,\psi_{r,p}\phi}\right) \qquad (2.37)
$$

$$
+ \sum_{q',w'}\bigg[\mathcal{D}_2(q,w,q',w')\,{}_{n'}h_{i,m'}^{x,\psi_{r,p}\psi_{q',w'}} + \mathcal{D}_3(q,w,q',w')\,{}_{n'}h_{i,m'-1}^{x,\psi_{r,p}\psi_{q',w'}}\bigg]
$$

$$
- \bigg[2^{r/2}\delta_{p,0}\left({}_{n'}h_{i',m}^{z,\phi\psi_{q,w}} - {}_{n'}h_{i'-1,m}^{z,\phi\psi_{q,w}}\right)
$$

$$
+ \sum_{r',p'}\bigg[\mathcal{D}_2(r,p,r',p')\,{}_{n'}h_{i',m}^{z,\psi_{r',p'}\psi_{q,w}} + \mathcal{D}_3(r,p,r',p')\,{}_{n'}h_{i',m-1}^{z,\psi_{r',p'}\psi_{q,w}}\bigg]\bigg]\bigg\}
$$

H_x-update equations

$$
{}_{n'}h_{i,m'}^{x,\phi\phi} = {}_{n'-1}h_{i,m'}^{x,\phi\phi} + \frac{\Delta t \Delta x}{\mu \Delta z}\bigg\{ {}_nE_{i,m+1}^{y,\phi\phi} - {}_nE_{i,m}^{y,\phi\phi} + \sum_{0 \le r \le r_{z,\max}} 2^{r/2}\left({}_nE_{i,m+1}^{y,\phi\psi_{r,0}} - {}_nE_{i,m}^{y,\phi\psi_{r,0}}\right)\bigg\}
$$

$$
(2.38)
$$

$$
{}_{n'}h_{i,m'}^{x,\zeta\phi} = {}_{n'-1}h_{i,m'}^{x,\zeta\phi} + \frac{\Delta t \Delta x}{\mu \Delta z}\bigg\{ {}_nE_{i,m+1}^{y,\zeta\phi} - {}_nE_{i,m}^{y,\zeta\phi} + \sum_{0 \le r \le r_{z,\max}} 2^{r/2}\left({}_nE_{i,m+1}^{y,\zeta\psi_{r,0}} - {}_nE_{i,m}^{y,\zeta\psi_{r,0}}\right)\bigg\}
$$

$$
(2.39)
$$

$$
{}_{n'}h_{i,m'}^{x,\phi\eta} = {}_{n'-1}h_{i,m'}^{x,\phi\eta} + \frac{\Delta t \Delta x}{\mu \Delta z}\bigg\{ 2^{r_z/2}\delta_{p_z,2^{r_z}-1}\left({}_nE_{i,m}^{y,\phi\phi} - {}_nE_{i,m+1}^{y,\phi\phi}\right) + \sum_{r'_z,p'_z}\mathcal{D}_0(r_z,p_z,
$$

$$
r'_z,p'_z)\,{}_nE_{i,m+1}^{y,\phi\psi_{r'_z,p'_z}} + \sum_{r'_z,p'_z}\mathcal{D}_1(r_z,p_z,r'_z,p'_z)\,{}_nE_{i,m}^{y,\phi\psi_{r'_z,p'_z}}\bigg\}
$$

$$
(2.40)
$$

$$
{}_{n'}h_{i,m'}^{x,\zeta\eta} = {}_{n'-1}h_{i,m'}^{x,\zeta\eta} + \frac{\Delta t \Delta x}{\mu \Delta z}\bigg\{ 2^{r_z/2}\delta_{p_z,2^{r_z}-1}\left({}_nE_{i,m}^{y,\zeta\phi} - {}_nE_{i,m+1}^{y,\zeta\phi}\right) + \sum_{r'_z,p'_z}\mathcal{D}_0(r_z,p_z,
$$

$$
r'_z,p'_z)\,{}_nE_{i,m+1}^{y,\zeta\psi_{r'_z,p'_z}} + \sum_{r'_z,p'_z}\mathcal{D}_1(r_z,p_z,r'_z,p'_z)\,{}_nE_{i,m}^{y,\zeta\psi_{r'_z,p'_z}}\bigg\}
$$

$$
(2.41)
$$

H_z-update equations

$$_{n'}h_{i',m}^{z,\phi\phi} = {}_{n'-1}h_{i',m}^{z,\phi\phi} - \frac{\Delta t \Delta z}{\mu \Delta x}\left\{ {}_nE_{i+1,m}^{y,\phi\phi} - {}_nE_{i,m}^{y,\phi\phi} + \sum_{0 \le r \le r_{x,\max}} 2^{r/2}\left({}_nE_{i+1,m}^{y,\psi_{r,0}\phi} - {}_nE_{i,m}^{y,\psi_{r,0}\phi} \right) \right\}$$

$$(2.42)$$

$$_{n'}h_{i',m}^{z,\phi\eta} = {}_{n'-1}h_{i',m}^{z,\phi\eta} - \frac{\Delta t \Delta z}{\mu \Delta x}\left\{ {}_nE_{i+1,m}^{y,\phi\eta} - {}_nE_{i,m}^{y,\phi\eta} + \sum_{0 \le r \le r_{x,\max}} 2^{r/2}\left({}_nE_{i+1,m}^{y,\psi_{r,0}\eta} - {}_nE_{i,m}^{y,\psi_{r,0}\eta} \right) \right\}$$

$$(2.43)$$

$$_{n'}h_{i',m}^{z,\zeta\phi} = {}_{n'-1}h_{i',m}^{z,\zeta\phi} - \frac{\Delta t \Delta z}{\mu \Delta x}\left\{ 2^{r_x/2}\delta_{p_x,2^{r_x}-1}\left({}_nE_{i,m}^{y,\phi\phi} - {}_nE_{i+1,m}^{y,\phi\phi} \right) + \sum_{r'_x,p'_x} \mathcal{D}_0\left(r_x, p_x, \right.\right.$$
$$\left.\left. r'_x, p'_x \right) {}_nE_{i+1,m}^{y,\psi_{r'_x,p'_x}\phi} + \sum_{r'_x,p'_x} \mathcal{D}_1(r_x, p_x, r'_x, p'_x) {}_nE_{i,m}^{y,\psi_{r'_x,p'_x}\phi} \right\}$$

$$(2.44)$$

$$_{n'}h_{i',m}^{z,\zeta\eta} = {}_{n'-1}h_{i',m}^{z,\zeta\eta} - \frac{\Delta t \Delta z}{\mu \Delta x}\left\{ 2^{r_x/2}\delta_{p_x,2^{r_x}-1}\left({}_nE_{i,m}^{y,\phi\eta} - {}_nE_{i+1,m}^{y,\phi\eta} \right) + \sum_{r'_x,p'_x} \mathcal{D}_0\left(r_x, p_x, \right.\right.$$
$$\left.\left. r'_x, p'_x \right) {}_nE_{i+1,m}^{y,\psi_{r'_x,p'_x}\eta} + \sum_{r'_x,p'_x} \mathcal{D}_1(r_x, p_x, r'_x, p'_x) {}_nE_{i,m}^{y,\psi_{r'_x,p'_x}\eta} \right\}$$

$$(2.45)$$

A numerical dispersion analysis of this system of equations yields the following expression [38]:

$$\left\{ \frac{1}{u_p \Delta t} \sin \frac{\omega \Delta t}{2} \right\}^2 = \left\{ \frac{1}{\Delta x_{\text{eff}}} \sin \frac{k_x \Delta x_{\text{eff}}}{2} \right\}^2$$
$$+ \left\{ \frac{1}{\Delta z_{\text{eff}}} \sin \frac{k_z \Delta z_{\text{eff}}}{2} \right\}^2,$$

$$(2.46)$$

For completeness, the well-known FDTD equations for TE^z waves are provided below. An FDTD cell size of $\delta x \times \delta z$ and time step δt is assumed. Then:

$$_{n+1}E_{i,m}^{y} = {}_nE_{i,m}^{y} + \frac{\delta t}{\epsilon \delta x \delta z}\left\{ {}_{n+\frac{1}{2}}h_{i,m+\frac{1}{2}}^{x} - {}_{n+\frac{1}{2}}h_{i,m-\frac{1}{2}}^{x} - {}_{n+\frac{1}{2}}h_{i+\frac{1}{2},m}^{z} + {}_{n+\frac{1}{2}}h_{i-\frac{1}{2},m}^{z} \right\}$$

$$_{n+\frac{1}{2}}h_{i,m+\frac{1}{2}}^{x} = {}_{n-\frac{1}{2}}h_{i,m+\frac{1}{2}}^{x} + \frac{\delta t \delta x}{\mu \delta z}\left\{ {}_nE_{i,m+1}^{y} - {}_nE_{i,m}^{y} \right\}$$

$$_{n+\frac{1}{2}}h_{i+\frac{1}{2},m}^{z} = {}_{n-\frac{1}{2}}h_{i+\frac{1}{2},m}^{z} - \frac{\delta t \delta z}{\mu \delta x}\left\{ {}_nE_{i+1,m}^{y} - {}_nE_{i,m}^{y} \right\}$$

$$(2.47)$$

A dispersion analysis for the system of FDTD equations (2.47), leads to the following expression [39]:

$$\left\{\frac{1}{u_p\delta t}\sin\frac{\omega\delta t}{2}\right\}^2 = \left\{\frac{1}{\delta x}\sin\frac{k_x\,\delta x}{2}\right\}^2 + \left\{\frac{1}{\delta z}\sin\frac{k_z\,\delta z}{2}\right\}^2. \tag{2.48}$$

By inspection of Eq. (2.46) and (2.48), it is readily concluded that under the condition:

$$\Delta x_{\text{eff}} \equiv \delta x, \quad \Delta z_{\text{eff}} \equiv \delta z, \tag{2.49}$$

the FDTD and Haar MRTD schemes present the same dispersion properties. Hence, a reflection-less interface between the two methods can be established. Time marching is carried out concurrently in the FDTD and MRTD regions, since the same time step is chosen for both methods. The common dispersion equation of the two schemes also implies common stability properties. As a result, a stable time step for the MRTD region is also stable for FDTD.

2.4.2   Connection Algorithm

A simple, one-dimensional example of an interface between a first-order Haar MRTD and the corresponding FDTD scheme is shown in Fig. 2.13. For the update of the electric field MRTD coefficients E_5^ϕ, E_5^ψ, $E_5^{\psi_{10}}$, $E_5^{\psi_{11}}$, magnetic scaling and wavelet terms one cell to the left, within the FDTD region, are necessary. For their calculation, the FDTD nodal values of the magnetic field H_1, H_2, H_3, H_4 are used as an input to a recursive fast wavelet transform (FWT). Thus, the wavelet decomposition of the magnetic field at that cell is deduced and employed for the electric field updates of MRTD, at the FDTD/MRTD boundary. In the case of an FDTD to MRTD transition, an inverse fast wavelet transform (IFWT) is applied, to derive FDTD coefficients (nodal field values) from MRTD terms. It is noted that both FWT and IFWT are characterized by an optimal complexity of $\mathcal{O}(N)$. In addition, their computational implementation is relatively simple.

As a computational validation of these interface principles, the propagation of a 0–5 GHz Gaussian pulse through an FDTD/fourth-order MRTD interface is shown in Fig. 2.14. The cell size for the FDTD region (and *effective* cell size for MRTD) is 2.4 mm, while the MRTD scaling cell size is 76.8 mm. The time step is 0.9 of the Courant limit. The interface for the electric field nodes is located at cell 1440. Smooth and stable transition from the MRTD to FDTD region is observed, as expected.

In the following, a two-dimensional FDTD/MRTD connection algorithm is presented, for the case where an FDTD region encloses an MRTD one. An application of interest is the termination of an MRTD mesh in an FDTD perfectly matched layer absorber, that allows for MRTD domain truncation via existing absorber codes. Similar concepts can be employed for interfaces of other types.

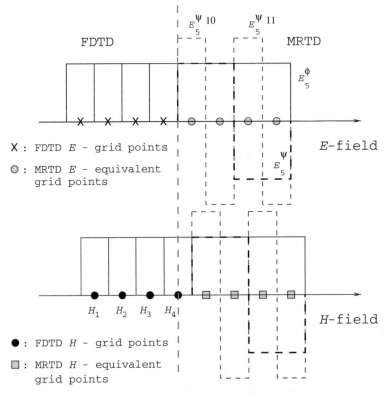

FIGURE 2.13: One-dimensional interface between FDTD and a first-order Haar MRTD scheme

To begin with, data transfer from MRTD to FDTD is addressed. For the update of all FDTD grid points whose stencil extends into the MRTD region, it suffices to retrieve the nodal field values of the electric field one FDTD cell within the MRTD region (Fig. 2.15). Thus, the complete determination of the tangential electromagnetic field components along the boundary of the FDTD domain is allowed for. Then, the independent solution of this region becomes possible, since the MRTD calculated tangential fields are used as boundary conditions for FDTD. In consequence, the problem is reduced to calculating nodal field values along the boundary of the MRTD region, which is exactly the function of IFWT.

Conversely, for the update of the MRTD boundary magnetic field coefficients, FDTD electric field nodes extending over one scaling cell within the FDTD region are wavelet transformed, via a two-dimensional FWT routine. If this scaling cell extends beyond the domain, zero field values are used. This is possible and physically correct for both closed and open domain problems, since a perfect conductor-backed absorber is typically used for the simulation of an open boundary. Figure 2.16 schematically explains this procedure, for a case where $r_{x,\max} = r_{z,\max} = 1$. It is noted that no FDTD grid points are sought for at the (shaded) corner

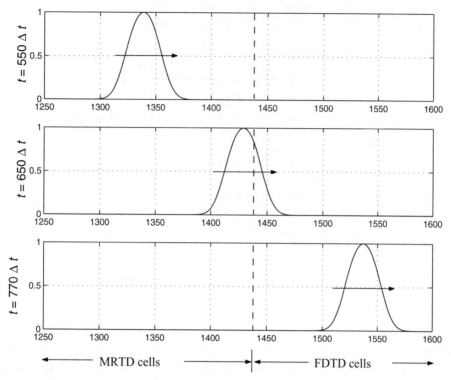

FIGURE 2.14: Reflection-less propagation through a one-dimensional FDTD/fourth-order MRTD interface

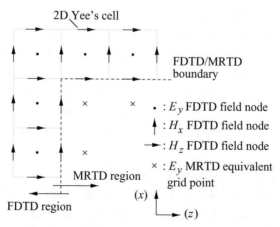

FIGURE 2.15: FDTD update from MRTD data

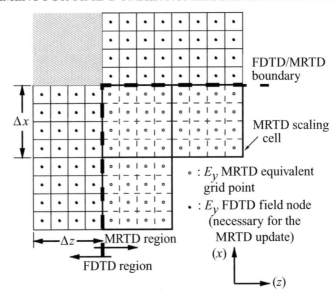

FIGURE 2.16: MRTD update from FDTD data

region, shown in Fig. 2.16. This is due to the fact that MRTD, just as FDTD, uses a cross-shaped stencil for the update of all grid points. Computing the electric field MRTD coefficients from the FDTD data via an FWT and updating the tangential magnetic field component coefficients via the standard MRTD finite difference equations, determines again the tangential electromagnetic field components along the boundary of the MRTD region. This, in turn, is sufficient for its independent, MRTD-based solution.

Evidently, all operations that implement this connection algorithm are performed at the same time step, during the update of the electric field coefficients in both regions, in an absolutely stable fashion, due to the matching of the dispersion properties of the two schemes. Furthermore, the same principles lead to interfaces between wavelet schemes of an arbitrary basis and the ones that are formulated by the corresponding scaling functions only, provided that the effective resolutions in the two regions are kept the same. It is also noted that the extension of the interface algorithm to three dimensions is accomplished by treating each face of Yee's cell according to the method that has been set forth in this work.

2.5   NUMERICAL RESULTS: VALIDATION

Several two-dimensional air-filled square resonators have been analyzed in various FDTD/MRTD mesh configurations for validation purposes. The choice of those configurations was made in order to demonstrate that the stability and accuracy of the algorithm was preserved under any gridding conditions, these being either related to the order of the MRTD

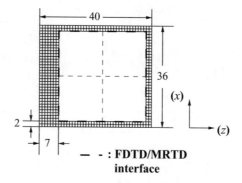

FIGURE 2.17: Empty rectangular cavity geometry and interface of a 3- by 3-order MRTD/FDTD mesh configuration (dimensions are given in FDTD cells, MRTD mesh is 2 × 2)

scheme or the proximity of the MRTD region to the hard boundary of the domain. The dimensions of the cavity structures in all figures are given in FDTD cells (or MRTD *equivalent* grid points). Figures 2.17 and 2.18 depict two case studies for MRTD/FDTD mesh configurations. In the first case (Fig. 2.17), an MRTD scheme of order 3 in both x- and z-directions forms a mesh of 2 by 2 scaling cells corresponding to 32×32 FDTD cells, asymmetrically placed within an FDTD region, that is terminated into the metal boundaries of the cavity. The whole domain corresponds to 36×40 FDTD cells of dimension 1 cm × 1 cm. A pure MRTD scheme would model these electric walls by means of image theory, necessitating the introduction of several images of high-order wavelet terms [11]. On the other hand, enclosing the MRTD region in an FDTD one, facilitates the treatment of these hard boundaries, whose FDTD modeling amounts to setting the tangential to PEC electric field nodal values equal to zero. Similarly, in the second case, an MRTD scheme of orders 3 and 4 in x- and z-directions respectively forms a 1 by 1 scaling mesh and is interfaced with FDTD of cell size 1 cm × 1 cm.

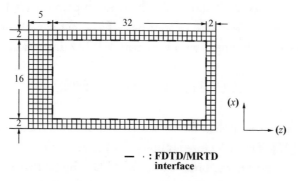

FIGURE 2.18: Empty rectangular cavity geometry and interface of a 3- by 4-order MRTD/FDTD mesh configuration (dimensions are given in FDTD cells, MRTD mesh is 1 × 1)

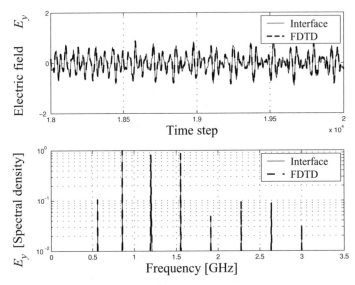

FIGURE 2.19: Time and frequency domain patterns of electric field (E_y) sampled within the cavity of Fig. 2.17, as obtained by FDTD and the hybrid scheme

The total domain corresponds to 20×39 FDTD cells. Cavity resonances that are derived via a pure FDTD scheme and the proposed interface-based method agree well (in both time and frequency domain), as shown in Figs. 2.19 and 2.20.

In all cases the time step was set equal to 0.9 of the Courant stability limit.

A third validation case is shown in Fig. 2.21. In that case, the MRTD region is symmetrically placed within the FDTD mesh and only two FDTD cells away from the boundary. The MRTD scheme is of order 4 in both directions, hence giving rise to a single scaling cell mesh in the MRTD region. The effective cell size is again 1 cm $\times$ 1 cm and the time step equal to 0.9 of the Courant limit. The patterns of the TE_{21}- and TE_{22}-modes, derived via the interface algorithm, are shown in Figs. 2.22 and 2.23. The smooth patterns confirm the correctness of the scheme and the absence of any type of spurious effects that would corrupt its performance. This observation is also in agreement with the Haar MRTD dispersion analysis of [38].

2.6   NUMERICAL RESULTS: APPLICATIONS

2.6.1   Metal Fin-Loaded Cavity

The method of this chapter is applied for the simulation of a metal fin-loaded cavity, similar to the one presented in [27]. This structure is chosen for the reason that the presence of the metal fin within the domain, restricts the order of the MRTD scheme that can be employed for its analysis. In particular, whenever a scaling cell greater than the fin dimensions is chosen, utmost care is necessary for the compensation of the unphysical coupling of the regions below and above the fin, caused by the scaling function defining the fin cell (or wavelets extending beyond the fin

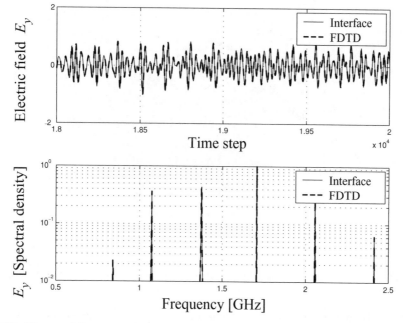

FIGURE 2.20: Time and frequency domain patterns of electric field (E_y) sampled within the cavity of Fig. 2.18, as obtained by FDTD and the hybrid scheme

limits). However, the strategies that are followed in this case (e.g., domain split [40]), result in a local increase of operations and consumption of computational resources. Furthermore, they bring about practical implementation problems, especially when one is interested in developing generic wavelet-based CAD tools.

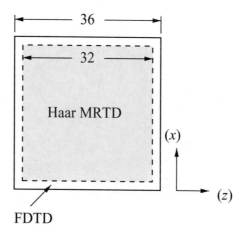

FIGURE 2.21: Empty square cavity geometry and interface of a 4- by 4-order MRTD/FDTD mesh configuration (dimensions are given in FDTD cells, MRTD mesh is 1×1)

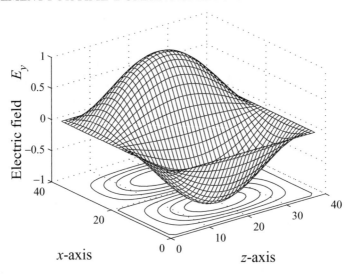

FIGURE 2.22: Electric field pattern for the TE_{22}-mode of the empty cavity of Fig. 2.21 (4 by 4 MRTD)

For the interface-based solution of the problem, the gridding conditions are given in Fig. 2.24. In the MRTD region, a second- by second-order scheme is employed (4 by 4 scaling cells). The time step is set at 0.8 of the Courant limit and FDTD cells of 1 cm × 1 cm are used. As an excitation, a Gaussian pulse, with its 3 dB frequency chosen to be equal to $f_c/2$, f_c being the cut-off frequency of the TE_{11} mode of the cavity, is applied in the FDTD region (at the plane $z = 5$ cm). Under these conditions, an absolutely stable performance of the code

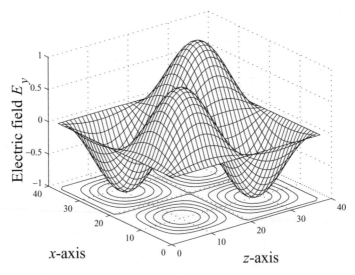

FIGURE 2.23: Electric field pattern for the TE_{22}-mode of the empty cavity of Fig. 2.21 (4 by 4 MRTD)

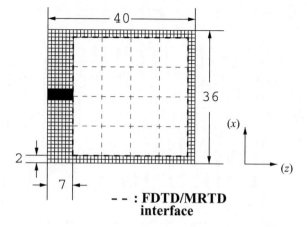

FIGURE 2.24: Metal fin-loaded cavity geometry and interface of a 2- by 2-order MRTD/FDTD mesh configuration (dimensions are given in FDTD cells, MRTD mesh is 4 × 4)

was obtained. The deduced electric field spatial distribution is shown in Fig. 2.25. Moreover, in order to demonstrate the stability of the solution, the electric field, sampled at the point ($x = 1.55$ cm, $z = 2.35$ cm), is plotted as a function of time for an arbitrary interval of 18,000–20,000 time steps, in Fig. 2.26. It is noted that no late-time instabilities were observed over as many as 100,000 time steps.

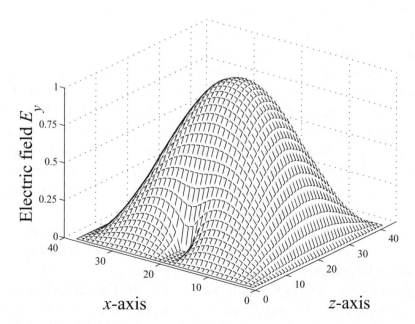

FIGURE 2.25: Electric field distribution in the metal fin-loaded cavity (dominant TE-mode)

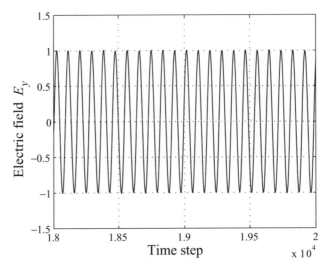

FIGURE 2.26: Electric field sampled within the fin-loaded cavity as a function of time for time steps 18,000–20,000.

2.6.2   Mesh Truncation

2.6.2.1 PML Absorber

A typical problem related to higher-order and multiresolution formulations of finite difference schemes is the inherently complex modeling of boundary and material conditions. For high-order Haar wavelet schemes, a formulation for a matched layer absorber that was proposed in [41] highlighted the difficulty of such efforts, that essentially stems from the sampling of varying electric and magnetic conductivities by a multilevel basis. On the other hand, in [37], the approach of omitting wavelet operations in the absorber region was adopted. However, the obvious compromise in accuracy that is related to this approach has negligible effect only in radiation problems, under the condition that the absorber has been placed far enough from the source. In this case, field wavelet coefficients at the absorber are small enough and the numerical error produced by their omission is controllable. On the contrary, this method is not applicable to open guiding structures because of the significant field values that typically impinge upon the absorber region of their computational domain.

Furthermore, it may seem necessary that the absorber region of an MRTD mesh be discretized by at least the degrees of freedom of a single MRTD cell. However, for high-order schemes, this would lead to absorber regions that are much longer than the ones typically used in FDTD applications. For example, terminating an MRTD with five wavelet levels (fourth-order scheme) in just a single cell of it, is equivalent to using an FDTD absorber of 64 grid points per direction. Nevertheless, excellent performance of 8–16 cell, optimized FDTD-uniaxial perfectly matched layer (UPML) absorbers has been recently demonstrated [42], prompting the quest for MRTD absorbers that may extend over a *fraction* of an MRTD scaling cell.

In this work, the concept of the FDTD/MRTD interface is employed for the implementation of a UPML termination of an MRTD domain. In particular, an FDTD-UPML region encloses an MRTD one and the interface algorithm is applied for the reflection-less connection of the two. In case the FDTD region corresponds to a fraction of a scaling cell, the application of MRTD update equations needs FDTD grid points beyond the conductor that backs the PML. The latter are simply zeroed out and fed back as such to the FWT routine. This concept is explained for a one-dimensional case of a second-order MRTD scheme (with eight equivalent grid points per cell), terminated into a four-cell FDTD PML (half a scaling cell), in Fig. 2.27.

Using this method, the waveguide structure of Fig. 2.28, also presented in [17], is solved by a fourth-order MRTD scheme, truncated with a 6 and 8 grid point PML corresponding to 0.1875 and 0.25 of a scaling cell, which in this case is 8 mm. A 0–30 GHz Gaussian pulse excitation is used, and the reflection coefficient from the slab is calculated. Hence, the scaling cell is $1.28 \times \lambda_{min}$, while the five wavelet levels successively refine the resolution of the scheme to $\lambda_{min}/25$. In Fig. 2.29, waveforms of incident and reflected electric fields, analyzed in their scaling and wavelet constituents are shown. Their correctness is readily verified, by noticing that they assume the form of spatial field derivatives, as expected. Additionally, Fig. 2.30 depicts

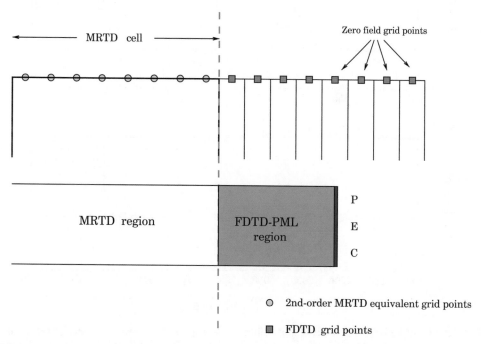

FIGURE 2.27: Concept of MRTD mesh termination via an FDTD/MRTD interface

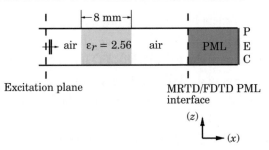

FIGURE 2.28: Slab waveguide geometry

comparative plots of the numerical results derived by the two termination types, along with the theoretical S_{11} form derived by transmission line theory. Evidently, all three sets of results are in good correlation with each other.

Then, a two-dimensional, 8 cell-UPML with theoretical reflection coefficient $R = \exp(-16)$ and fourth-order polynomial conductivity variation is used to terminate MRTD meshes that correspond to a 64×64 FDTD domain (Fig. 2.31). The three case studies are

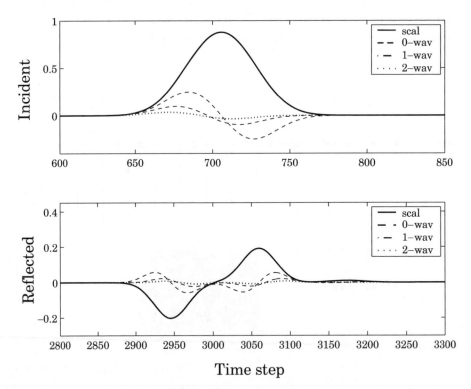

FIGURE 2.29: Waveforms of incident and reflected electric field (scaling and wavelet terms of order 0–2), sampled in front of the dielectric slab of Fig. 2.28

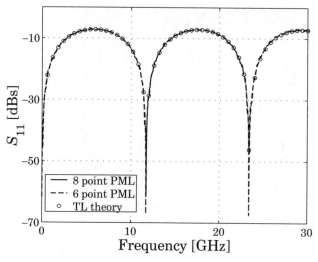

FIGURE 2.30: Numerical and theoretical S_{11} for the slab geometry

depicted in Figs. 2.32–2.34 and correspond to MRTD orders 0 by 0, 2 by 2, and 4 by 4, respectively. An electric current excitation of the form:

$$J_y(t) = \frac{1}{\Delta x\,\Delta z} \left(4\,(t/t_0)^3 - (t/t_0)^4\right)\,\exp(-t/t_0) \qquad (2.50)$$

with $t_0 = 1/(4\pi f_0)$ and $f_0 =$ 1 GHz, is applied in the middle of the domain. The FDTD cell size is set equal to 2.5 mm × 2.5 mm and the time step is $\Delta t = 4.5\,ps$. Time-domain field waveforms are then sampled at the points indicated as A, B, C in the three figures, corresponding

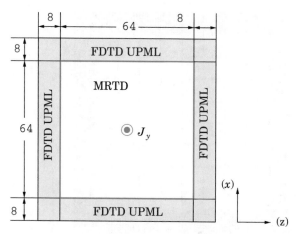

FIGURE 2.31: Case study for the interface-based termination of a two-dimensional MRTD domain

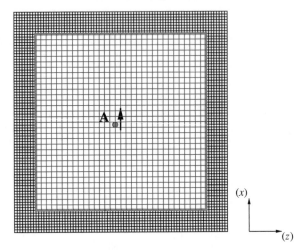

FIGURE 2.32: FDTD-UPML interfaced with a 0- by 0-order MRTD (MRTD mesh is 32 × 32)

to FDTD cells (31, 31), (17, 17), (1, 1) of the 64 × 64 domain. The results, shown in Fig. 2.35, demonstrate an excellent agreement between the pure FDTD scheme and the FDTD-UPML terminated MRTD. It is noted that the UPML regions in these three examples extend over 4, 1, and 0.25 MRTD cells respectively, demonstrating the ability of the interface to provide MRTD absorbing boundary conditions with efficiency and complexity that are independent of the order of the underlying wavelet expansion.

In addition, Fig. 2.36 depicts the electric field waveform sampled at point B of the MRTD domain of Fig. 2.33, when terminated at 4, 8, and 16 FDTD-UPML cells (0.5, 1, and 2 MRTD cells respectively). The same waveform is determined via the FDTD method, applied

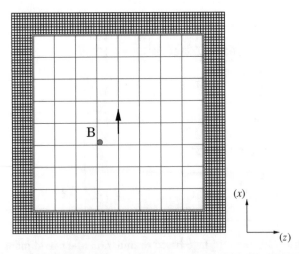

FIGURE 2.33: FDTD-UPML interfaced with a 2- by 2-order MRTD (MRTD mesh is 8 × 8)

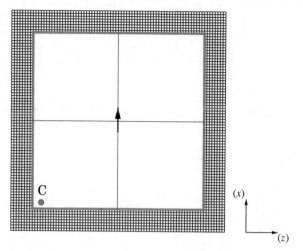

FIGURE 2.34: FDTD-UPML interfaced with a 4- by 4-order MRTD (MRTD mesh is 2×2)

at a 64×64 mesh, terminated at 32 UPML cells. Evidently, all four curves agree well. Moreover, the broad time window over which the results are given, shows the absence of any significant retro-reflections from the absorbers in all three termination schemes. Finally, Figs. 2.37 and 2.38 depict the spatial evolution of the pulse, its propagation toward and its absorption from the

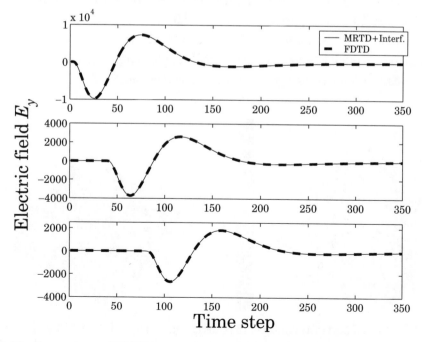

FIGURE 2.35: Comparison of FDTD and MRTD/FDTD interface results for the three case studies of Figs. 2.32–2.34

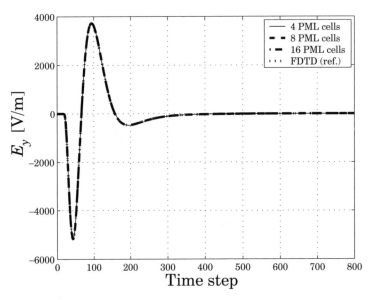

FIGURE 2.36: Electric field at point B of Fig. 2.33, when a 4, 8, 16 FDTD UPML is used to terminate the MRTD mesh. An FDTD solution (with a 32-cell UPML) is appended for comparison

FDTD-UPML boundaries, that are interfaced to the MRTD domain. In fact, the spatial field distribution at time step 600, essentially represents the level of errors due to artificial reflections from the absorber. This level appears to be at -80 dB.

In conclusion, one can consider the proposed method as an effective alternative for MRTD mesh termination, since it employs existing UPML implementations, linking them to MRTD codes via the simple connection algorithm that was earlier described.

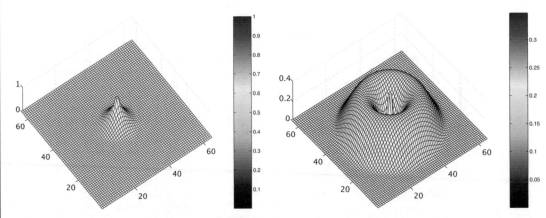

FIGURE 2.37: Spatial field distribution in the interface-terminated domain of Fig. 2.31 at time steps 20, 50

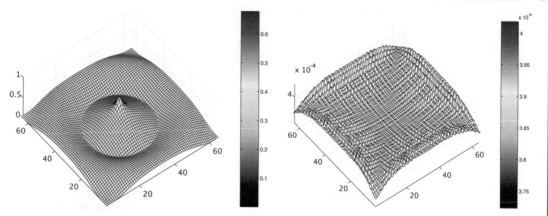

FIGURE 2.38: Spatial field distribution in the interface-terminated domain of Fig. 2.31 at time steps 80, 600

2.6.2.2 Absorbing Boundary Conditions

Evidently, even the application of simple absorbing boundary conditions (ABCs) assumes significant complexity in the context of multiresolution techniques. The reason for that is that ABCs impose mathematical conditions at planes that include only a fraction of the equivalent grid points contained within a terminal cell. The concept of an interface-based solution to this question is depicted in Fig. 2.39. For the purpose of applying an absorbing boundary condition, an FDTD region—no shorter than a single MRTD scaling cell—is introduced. Then, FDTD region is terminated into a classically implemented Mur's ABC. Thus, while the boundary condition itself takes a single cell, now the FDTD region needs to extend over the degrees of freedom of an MRTD scaling cell—not a fraction of it, as in the case of a PEC-backed PML absorber. Because of this trade-off, this mesh truncation method is more suitable to low-order MRTD schemes.

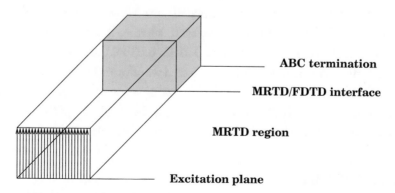

FIGURE 2.39: Concept of ABC termination of an MRTD mesh; the FDTD region terminated into the ABC should at least extend over one scaling MRTD cell

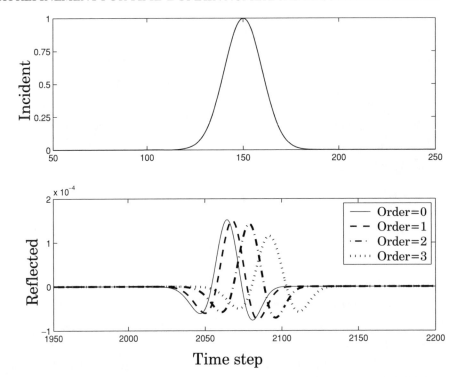

FIGURE 2.40: Incident and reflected field waveforms, for Mur's first-order ABC (implemented via an FDTD/MRTD interface), in the case of one-dimensional wave propagation (Fig. 2.39), for MRTD orders 0–3

Results from a one-dimensional implementation of these concepts are shown in Figs. 2.40 and 2.41. A 0–5 GHz Gaussian pulse propagation, in a TEM waveguide mode, is considered. Four Haar MRTD schemes of orders 0–3, with effective cell size of 0.0024 m and time step equal to 0.9 of their Courant limit, are terminated into Mur's first-order ABC. The FDTD region occupies two MRTD scaling cells for each scheme. Therefore, the thickness of the FDTD region for each case study is: 0.0096, 0.0192, 0.0384, 0.0768 m, respectively. That is also the reason for the phase difference of the reflected waveforms in Fig. 2.40. Finally, Fig. 2.41 presents the reflection coefficient of each of the four terminations. Obviously, the obtained ABC performance is similar in all MRTD cases and confirms the capability of the method to provide an efficient means of implementing an ABC-truncation of wavelet grids.

2.7   CONCLUSIONS

A numerical interface between an arbitrary-order Haar MRTD and FDTD was developed and applied in this chapter. The two unique features of the proposed technique are that first,

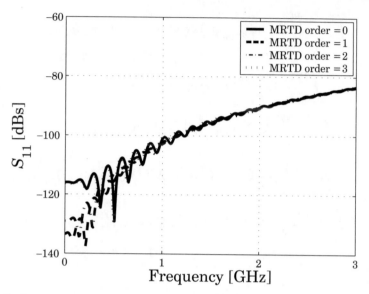

FIGURE 2.41: Reflection coefficient for Mur's first-order ABC (implemented via an FDTD/MRTD interface), in the case of one-dimensional wave propagation (Fig. 2.39)

it does not employ any interpolations or extrapolations and second, it is applicable for *any* MRTD order. Hence, the developed algorithm constitutes a computationally efficient tool for jointly exploiting the advantages of FDTD and MRTD, which is critical for the acceleration of time-domain schemes, when applied to large-scale problems of current microwave technology structures.

APPENDIX: COEFFICIENTS FOR HAAR MRTD UPDATE EQUATIONS

Analytical expressions of integrals that are encountered in the application of the method of moments for the derivation of Haar MRTD update equations are provided below. The maximum wavelet-order that determines the mutual shift of electric/magnetic field nodes in axial directions is denoted as R, while $s_R = 1/2^{R+2}$ is the shift itself.

$$\int_{-\infty}^{+\infty} \frac{d\phi_{k'}}{dt}(t)\phi_{k+s_R}(t)\, dt = \delta_{k',k+1} - \delta_{k',k} \tag{A.1}$$

$$\int_{-\infty}^{+\infty} \frac{d\psi_{k',p}^r}{dt}(t)\phi_{k+s_R}(t)\, dt = 2^{r/2}\left(\delta_{k',k+1} - \delta_{k',k}\right)\delta_{p,0} \tag{A.2}$$

$$\int_{-\infty}^{+\infty} \frac{d\phi_{k'}}{dt}(t)\psi_{k+s_R,p}^r(t)\, dt = 2^{r/2}\left(\delta_{k',k} - \delta_{k',k+1}\right)\delta_{p,2^r-1}$$

$$\int_{-\infty}^{+\infty} \frac{d\psi_{k',p'}^{r'}}{dt}(t)\psi_{k+s_R,p}^{r}(t)\,dt = \mathcal{D}_0(r,\,p,\,r',\,p')\delta_{k',k+1}$$
$$+\mathcal{D}_1(r,\,p,\,r',\,p')\delta_{k',k}$$

$$\int_{-\infty}^{+\infty} \frac{d\phi_{k'+s_R}}{dt}(t)\phi_k(t)\,dt = \delta_{k',k} - \delta_{k',k-1}$$

$$\int_{-\infty}^{+\infty} \frac{d\psi_{k'+s_R,p}^{r}}{dt}(t)\phi_k(t)\,dt = 2^{r/2}\left(\delta_{k',k-1} - \delta_{k',k}\right)\delta_{p,2^r-1}$$

$$\int_{-\infty}^{+\infty} \frac{d\phi_{k'+s_R}}{dt}(t)\psi_{k,p}^{r}(t)\,dt = 2^{r/2}\left(\delta_{k',k} - \delta_{k',k-1}\right)\delta_{p,0} \qquad (A.3)$$

$$\int_{-\infty}^{+\infty} \frac{d\psi_{k'+s_R,p'}^{r'}}{dt}(t)\psi_{k,p}^{r}(t)\,dt = \mathcal{D}_2(r,\,p,\,r',\,p')\delta_{k',k}$$
$$+\mathcal{D}_3(r,\,p,\,r',\,p')\delta_{k',k-1}$$

with

$$\mathcal{D}_0(r,\,p,\,r',\,p') = 2^{(r+r')/2}\left\{\psi(\xi_1) - 2\psi(\xi_2) + \psi(\xi_3)\right\}$$

$$\mathcal{D}_1(r,\,p,\,r',\,p') = 2^{(r+r')/2}\left\{\psi(\xi_1') - 2\psi(\xi_2') + \psi(\xi_3')\right\}$$

$$\mathcal{D}_2(r,\,p,\,r',\,p') = -\mathcal{D}_1(r,\,2^r - p - 1,\,r',\,2^{r'} - p' - 1) \qquad (A.4)$$

$$\mathcal{D}_3(r,\,p,\,r',\,p') = -\mathcal{D}_0(r,\,2^r - p - 1,\,r',\,2^{r'} - p' - 1)$$

and

$$\xi_1 = 2^r\left(1 + p'/2^{r'} - 1/2^{R+2}\right) - p$$
$$\xi_2 = 2^r\left(1 + (p' + 0.5)/2^{r'} - 1/2^{R+2}\right) - p$$
$$\xi_3 = 2^r\left(1 + (p' + 1)/2^{r'} - 1/2^{R+2}\right) - p$$
$$\xi_1' = 2^r\left(p'/2^{r'} - 1/2^{R+2}\right) - p \qquad (A.5)$$
$$\xi_2' = 2^r\left((p' + 0.5)/2^{r'} - 1/2^{R+2}\right) - p$$
$$\xi_3' = 2^r\left((p' + 1)/2^{r'} - 1/2^{R+2}\right) - p$$

Finally, $\delta_{\kappa,\lambda}$ is the well-known Kronecker delta and ψ the Haar mother wavelet function. The previous expressions are readily programmable and allow for the development of arbitrary-order Haar wavelet MRTD codes. Yet, code efficiency is greatly enhanced by *a priori* recognizing the *non-zero* coefficients $\mathcal{D}_0$, $\mathcal{D}_1$, $\mathcal{D}_2$, $\mathcal{D}_3$ and omitting operations that involve multiplications by zero in the main time-stepping loop. This is done at the pre-processing stage of an MRTD code.

CHAPTER 3

Efficient Implementation of Adaptive Mesh Refinement in the Haar Wavelet-based MRTD Technique

The main attractive feature of wavelet-based, time-domain techniques is the simple implementation of adaptive meshing, through the application of a thresholding procedure to eliminate wavelet coefficients that attain relatively insignificant values, at a limited compromise of accuracy. However, little attention has been devoted so far to the investigation of computational costs and accuracy trade-offs in order to obtain thresholding-related operation savings. This chapter presents an efficient implementation of thresholding applied to a nonlinear problem and reports significant execution time savings compared to the conventional FDTD technique, that the application of the proposed method has led to.

3.1 INTRODUCTION

Multiresolution analysis (MRA) concepts have already been employed in a wide range of applications [11,14,15,37]. A motivating force for this research activity is the fact that wavelet-based methods provide the most natural framework for the implementation of adaptive grids, dynamically following local variations and singularities of solutions to partial differential equations. In particular, it can be proven that the decay of wavelet expansion coefficients of a square integrable function depends on the local smoothness of the latter [32]. Hence, significant wavelet values are expected at space–time regions, where high variations in the numerical solution evolve. In this sense, sparing the arithmetic operations on wavelet coefficients below a certain threshold—small enough according to accuracy requirements—amounts to imposing coarse gridding conditions at those regions, while allowing for a denser mesh at parts of the domain where the solution varies less smoothly.

Several approaches to wavelet-based mesh refinement have been presented in the literature. In [37, 43], Haar wavelets were used to selectively refine the resolution of an underlying

FDTD scheme at parts of the domain where this seemed to be *a priori* necessary, in a static sense. However, the resulting method was strictly equivalent to a subgridded FDTD and presented obvious accuracy disadvantages, since it was not enhanced by the interpolatory operations, typically employed in subgridded FDTD [44]. In general, the necessity to incorporate wavelets into a time-domain simulation technique is always related to dynamic rather than static subgridding, since the latter—having been extensively studied in the literature—can be nowadays efficiently implemented by several existing engines.

On the contrary, adaptive, wavelet-based meshing was introduced in [32] and applied to electromagnetic structures in [12–15], always in conjunction with high-order basis functions, these being either Daubechies or B-splines. In both cases, the complexity of the proposed scheme made clear that "...the construction of wavelet-based discretizations with a robustness and flexibility comparable to FDTD is still a challenging task" [15]. The current work meets this challenge by building up a two-level scheme on the simplest wavelet basis, the Haar basis and attempting a simple and explicit implementation of an algorithm for the thresholding of wavelet coefficients, based on ideas originally related to shock-wave problems of computational fluid dynamics. Execution time measurements for the algorithm as applied to a nonlinear optics problem, show that this procedure can actually lead to faster-than-FDTD simulations.

3.2   WAVELET-BASED FRONT-TRACKING

In this section, the case study of the incidence of a 0–30 GHz Gaussian pulse on a dielectric slab (Fig. 2.28, Section 2.6.2.1) is revisited. Thresholding of Haar wavelet coefficients is applied, based on the condition:

$$\left| {}_nE_j^{\psi_{r,p}} \right| \leq 2^{-r/2}\epsilon,$$

where ϵ is a predefined threshold. Hence, the amplitude of the threshold is adapted to the order r of each wavelet coefficient, through the use of a normalization term $2^{-r/2}$. This choice is due to the fact that the maximum absolute value of the basis function $\psi_{j,p}^r(x)$ is $2^{r/2}$. Then, thresholding amounts to omitting wavelet terms whose maximum contribution to the field value is less than ϵ. For two different threshold values, $\epsilon = 10^{-6}$ and 10^{-4}, the number of wavelet coefficients that are above threshold is computed at each time step. The resultant plots, for wavelets of orders 0–3 are shown in Figs. 3.1 and 3.2. Evidently, despite small differences between them, all four curves have a similar pattern: Around time step 2000, unthresholded wavelet coefficients are doubled, as a result of the pulse incidence on the dielectric slab, that generates an additional (reflected) wavefront. Then, the PML absorption of the two wavefronts, symmetrically generated by the source in the middle of the domain, causes a stepwise decrease

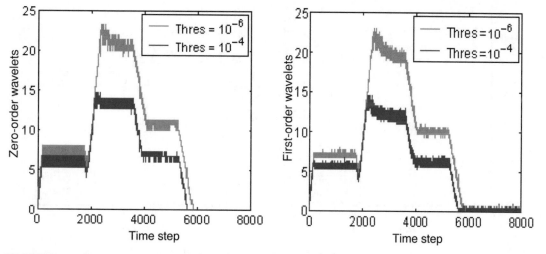

FIGURE 3.1: Unthresholded to total number (%) of wavelet coefficients of orders 0, 1, for a fourth-order Haar MRTD simulation of the problem of Fig. 2.28

in the number of active wavelet coefficients. Finally, the absorption of the reflected wavefront signals the end of the simulation and the decay of the number of unthresholded coefficients to almost zero. Overall, thresholding of wavelet coefficients yields a compression in memory requirements by 64.6% (in 8192 time steps). Also, Fig. 3.3 depicts the reflection coefficient

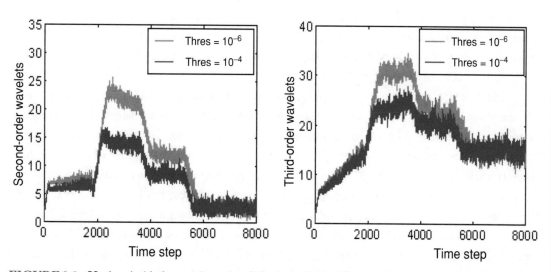

FIGURE 3.2: Unthresholded to total number (%) of wavelet coefficients of orders 2, 3, for a fourth-order Haar MRTD simulation of the problem of Fig. 2.28

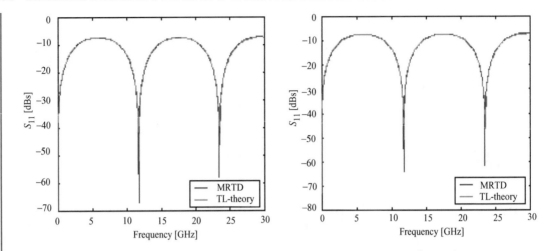

FIGURE 3.3: Comparison of S_{11}, derived by adaptive MRTD with $\epsilon = 10^{-5}$, 10^{-4} for the dielectric slab incidence problem of Fig. 2.28

S_{11} for the dielectric slab, as computed via adaptive MRTD. The excellent comparison of the numerical results to transmission line theory, demonstrates that wavelet thresholding has a limited effect on the accuracy of the algorithm.

Yet, this aspect of wavelet adaptivity will not be further investigated in this work, because taking advantage of wavelet compression at runtime, presupposes the application of memory operations (re-shuffle of field arrays) that have an adverse effect on execution time. It is noted though, that there are applications, where field values at all time steps for a certain domain need to be stored in memory. *A posteriori* processing those and wavelet thresholding them, can lead to significant memory economy [45].

Moreover, Figs. 3.4 and 3.5 depict six snapshots of electric field distribution in space, for the same problem, solved with a one-level Haar MRTD scheme (under same gridding conditions as before). The source is located in the middle of the domain and produces two symmetric, left- and right-propagating waves. At each time step, the regions of unthresholded wavelet coefficients (for $\epsilon = 10^{-4}$) are found and their boundaries are plotted (in blue dots). Two observations are now in order: First, wavelets follow successfully the field wavefronts, throughout the domain. Second, in order to keep track of the evolution of wavelet coefficients, one has to essentially study the movement of the boundaries of the unthresholded wavelet regions. This task is limited to tracking the *direction* of the velocity of those boundaries. Indeed, one can interpret the Courant stability condition ($c\,\Delta t \le \Delta x$), as follows: Wavefront position can change at each time step *at most* by one cell. In the following, these observations are exploited for the efficient application of thresholding in MRTD.

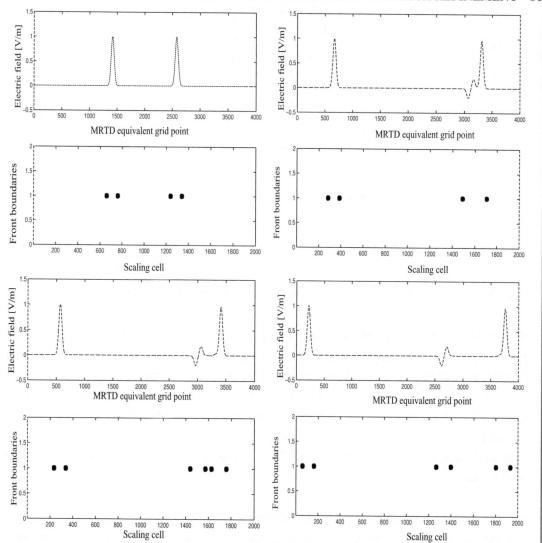

FIGURE 3.4: Evolution of wavefronts in the dielectric slab simulation (Fig. 2.28) and regions of unthresholded wavelet coefficients (boundaries are defined by blue dots, a threshold $\epsilon = 10^{-4}$ is used)

3.3   ADAPTIVE HAAR WAVELET SIMULATION OF PULSE COMPRESSION IN AN OPTICAL FIBER FILTER

As a vehicle for the demonstration of the algorithms developed in this study, the propagation of an optical pulse through a fiber filter and its gradual compression is simulated. This nonlinear phenomenon is based on a self-phase modulation (SPM) induced negative dispersion that the pulse experiences along the fiber and was originally reported and studied in [46]. It is noted that

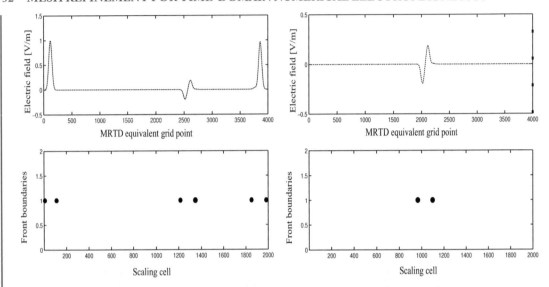

FIGURE 3.5: Evolution of wavefronts in the dielectric slab simulation (continued)

the refractive index in the fiber assumes the form: $n(z) = n_0 + n_1 \cos(2\beta_0 z) + n_2 |E|^2$, where the cosinusoidal term is due to a periodic structure (grating) written within the core of the fiber. Numerical modeling of this case was also pursued in [47], by means of the Battle–Lemarie cubic spline-based S-MRTD technique, resulting in execution time that was reported to be larger than FDTD by a factor of 1.5. In this work, a one wavelet-level Haar MRTD scheme is used, for the purpose of estimating the improvement in computational performance that exclusively originates from the dynamic adaptivity of MRTD rather than the high-order of an underlying scaling basis.

3.3.1   Formulation of Haar MRTD Scheme

The electric field in the fiber is decomposed in forward and backward propagating waves as:

$$E(z, t) = E_F(z, t)\, e^{j(\beta z - \omega t)} + E_B(z, t)\, e^{-j(\beta z + \omega t)}. \tag{3.1}$$

Substituting this expression into Maxwell's equations and discarding terms with spatial variation faster than $e^{j2\beta z}$ (slowly varying envelope approximation), the following system of equations is deduced:

$$\frac{\partial E_F}{\partial z} + \frac{n_0}{c}\frac{\partial E_F}{\partial t} = j\kappa E_B\, e^{-2j\Delta\beta z} + j\gamma \left(|E_F|^2 + 2|E_B|^2 \right) E_F \tag{3.2}$$

$$\frac{\partial E_B}{\partial z} - \frac{n_0}{c}\frac{\partial E_B}{\partial t} = -j\kappa E_F\, e^{2j\Delta\beta z} - j\gamma \left(|E_B|^2 + 2|E_F|^2 \right) E_B, \tag{3.3}$$

with $\gamma = \pi n_2/\lambda$, $\kappa = \pi n_1/\lambda$ and $\Delta\beta(\omega) = n_0(\omega)\omega/c - 2\pi n_0/\lambda_0$ and λ_0 the free-space wavelength that satisfies the Bragg condition for the grating.

The FDTD update equations for the system of (3.2), (3.3) can be retrieved from [47]. In this work, Haar MRTD equations are derived, by first expanding both forward and backward fields in terms of Haar scaling $\phi_m(z) = \phi(z/\Delta z - m)$ and wavelet $\psi_m(z) = \psi(z/\Delta z - m)$ functions in space and pulse functions $h_k(t) = h(t/\Delta t - k)$ in time. For these functions, definitions (2.4), (2.6), (2.14) are followed. Field expansions read:

$$E_x(z, t) = \sum_{m,k=-\infty}^{\infty} \left({}_kE_m^{x,\phi}\,\phi_m(z) + {}_kE_m^{x,\psi}\,\psi_m(z) \right) h_k(t), \qquad (3.4)$$

where $x = F, B$. Upon substitution of (3.4) into (3.2), (3.3) and application of the method of moments, the system of Haar MRTD update equations is formulated. In particular, update equations for the mth cell values of $E^{F,\phi}$, $E^{F,\psi}$, $E^{B,\phi}$, $E^{B,\psi}$ at the kth time step read:

$$
\begin{aligned}
{}_{k+1}E_m^{F,\,\phi} = {}_{k-1}E_m^{F,\,\phi} &- s\left\{ {}_kE_{m+1}^{F,\phi} - {}_kE_{m-1}^{F,\phi} + \left({}_kE_{m+1}^{F,\psi} - 2\,{}_kE_m^{F,\psi} + {}_kE_{m-1}^{F,\psi} \right) \right\} \\
&+ 2j\kappa s\,\Delta z e^{-2j\Delta\beta(m+1/2)\Delta z}\left\{ \mathbf{Sa}(\Delta\beta\Delta z)\,{}_kE_m^{B,\phi} + j\mathbf{Sa'}(\Delta\beta\Delta z/2)\,{}_kE_m^{B,\psi} \right\} \quad (3.5) \\
&+ 2j\gamma s\,\Delta z\left\{ {}_k\mathcal{A}_m\,{}_kE_m^{F,\phi} + {}_k\mathcal{B}_m\,{}_kE_m^{F,\psi} \right\},
\end{aligned}
$$

$$
\begin{aligned}
{}_{k+1}E_m^{F,\,\psi} = {}_{k-1}E_m^{F,\,\psi} &+ s\left\{ {}_kE_{m+1}^{F,\psi} - {}_kE_{m-1}^{F,\psi} + \left({}_kE_{m+1}^{F,\phi} - 2\,{}_kE_m^{F,\phi} + {}_kE_{m-1}^{F,\phi} \right) \right\} \\
&+ 2j\kappa s\,\Delta z e^{-2j\Delta\beta(m+1/2)\Delta z}\left\{ \mathbf{Sa}(\Delta\beta\Delta z)\,{}_kE_m^{B,\psi} + j\mathbf{Sa'}(\Delta\beta\Delta z/2)\,{}_kE_m^{B,\phi} \right\} \quad (3.6) \\
&+ 2j\gamma s\,\Delta z\left\{ {}_k\mathcal{A}_m\,{}_kE_m^{F,\psi} + {}_k\mathcal{B}_m\,{}_kE_m^{F,\phi} \right\},
\end{aligned}
$$

$$
\begin{aligned}
{}_{k+1}E_m^{B,\,\phi} = {}_{k-1}E_m^{B,\,\phi} &+ s\left\{ {}_kE_{m+1}^{B,\phi} - {}_kE_{m-1}^{B,\phi} + \left({}_kE_{m+1}^{B,\psi} - 2\,{}_kE_m^{B,\psi} + {}_kE_{m-1}^{B,\psi} \right) \right\} \\
&+ 2j\kappa s\,\Delta z e^{2j\Delta\beta(m+1/2)\Delta z}\left\{ \mathbf{Sa}(\Delta\beta\Delta z)\,{}_kE_m^{F,\phi} - j\mathbf{Sa'}(\Delta\beta\Delta z/2)\,{}_kE_m^{F,\psi} \right\} \quad (3.7) \\
&+ 2j\gamma s\,\Delta z\left\{ {}_k\mathcal{C}_m\,{}_kE_m^{B,\phi} + {}_k\mathcal{D}_m\,{}_kE_m^{B,\psi} \right\},
\end{aligned}
$$

$$
\begin{aligned}
{}_{k+1}E_m^{B,\,\psi} = {}_{k-1}E_m^{B,\,\psi} &- s\left\{ {}_kE_{m+1}^{B,\psi} - {}_kE_{m-1}^{B,\psi} + \left({}_kE_{m+1}^{B,\phi} - 2\,{}_kE_m^{B,\phi} + {}_kE_{m-1}^{B,\phi} \right) \right\} \\
&+ 2j\kappa s\,\Delta z e^{2j\Delta\beta(m+1/2)\Delta z}\left\{ \mathbf{Sa}(\Delta\beta\Delta z)\,{}_kE_m^{F,\psi} - j\mathbf{Sa'}(\Delta\beta\Delta z/2)\,{}_kE_m^{F,\phi} \right\} \quad (3.8) \\
&+ 2j\gamma s\,\Delta z\left\{ {}_k\mathcal{C}_m\,{}_kE_m^{B,\psi} + {}_k\mathcal{D}_m\,{}_kE_m^{B,\phi} \right\},
\end{aligned}
$$

with $\mathbf{Sa}(w) = \sin w/w$, $\mathbf{Sa'}(w) = \sin^2 w/w$ and $s = c\,\Delta t/n_0\Delta z$, where Δz is the MRTD scaling cell. Furthermore,

$$
_k\mathcal{A}_m = |\,_kE_m^{F,\phi}|^2 + |\,_kE_m^{F,\psi}|^2 + 2\left(|\,_kE_m^{B,\phi}|^2 + |\,_kE_m^{B,\psi}|^2\right),
$$

$$
_k\mathcal{B}_m = 1/2\left\{|\,_kE_m^{F,\phi} + \,_kE_m^{F,\psi}|^2 - |\,_kE_m^{F,\phi} - \,_kE_m^{F,\psi}|^2\right.
$$
$$
\left. + 2\left(|\,_kE_m^{B,\phi} + \,_kE_m^{B,\psi}|^2 - |\,_kE_m^{B,\phi} - \,_kE_m^{B,\psi}|^2\right)\right\},
$$

$$
_k\mathcal{C}_m = |\,_kE_m^{B,\phi}|^2 + |\,_kE_m^{B,\psi}|^2 + 2\left(|\,_kE_m^{F,\phi}|^2 + |\,_kE_m^{F,\psi}|^2\right), \tag{3.9}
$$

$$
_k\mathcal{D}_m = 1/2\left\{|\,_kE_m^{B,\phi} + \,_kE_m^{B,\psi}|^2 - |\,_kE_m^{B,\phi} - \,_kE_m^{B,\psi}|^2\right.
$$
$$
\left. + 2\left(|\,_kE_m^{F,\phi} + \,_kE_m^{F,\psi}|^2 - |\,_kE_m^{F,\phi} - \,_kE_m^{F,\psi}|^2\right)\right\}.
$$

It is worth mentioning that operations in (3.5)–(3.8) can be readily split in scaling- and wavelet-related ones, thus facilitating their adaptive application.

3.3.2   Boundary Conditions

Assuming an optical fiber that extends from $z = 0$ to $z = L$, the following boundary conditions are imposed [47]:

$$
E_F(0, t) = \sqrt{\frac{A}{2}}\,(1 + j)\,e^{-t^2/(2\alpha^2)}, \qquad E_B(L, t) = 0, \tag{3.10}
$$

by local modification of the update equations, based on the explicit enforcement of those conditions (without extrapolations or interpolations as proposed in the past [14]). For example, if $z = 0$ belongs to the mth scaling cell of the domain, and $g_k = E_F(0, k\Delta t)$ is the discrete-time sample of the excitation function at the kth step, applied at the Mth equivalent grid point (Fig. 3.6), then, by the wavelet transform:

$$
_kE_M^F = \frac{1}{\sqrt{2}}\left(\,_kE_m^{F,\phi} + \,_kE_m^{F,\psi}\right) = g_k/\sqrt{2}, \tag{3.11}
$$

$$
_kE_{M+1}^{F/B} = \frac{1}{\sqrt{2}}\left(\,_kE_m^{F/B,\phi} - \,_kE_m^{F/B,\psi}\right), \tag{3.12}
$$

$$
_kE_{M+2}^{F/B} = \frac{1}{\sqrt{2}}\left(\,_kE_{m+1}^{F/B,\phi} + \,_kE_{m+1}^{F/B,\psi}\right). \tag{3.13}
$$

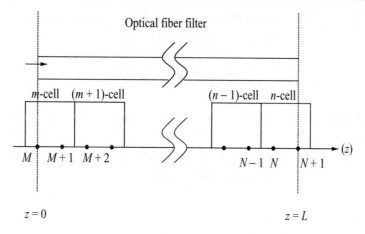

FIGURE 3.6: Haar MRTD equivalent grid points at the beginning and the end of the fiber

Defining an auxiliary variable: $A_k \equiv \sqrt{2}\ _k E_{M+1}^F$, the following, explicit update equation is derived:

$$
\begin{aligned}
A_{k+1} = A_{k-1} - 2s\left(\ _k E_{m+1}^{F,\phi} + \ _k E_{m+1}^{F,\psi} - g_k\right) \\
+ 2j\kappa s\,\Delta z \mathbf{Sa}\left(\Delta\beta\Delta z/2\right) e^{-j\Delta\beta(M+3/2)\Delta z}\left(\ _k E_m^{B,\phi} - \ _k E_m^{B,\psi}\right) \\
+ 2j\gamma s\,\Delta z \left\{\frac{1}{2}\left|\ _k E_m^{F,\phi} - \ _k E_m^{F,\psi}\right|^2 + \left|\ _k E_m^{B,\phi} - \ _k E_m^{B,\psi}\right|^2\right\}
\end{aligned}
\tag{3.14}
$$

Then, scaling and wavelet terms of the forward field are updated at the excitation cell, as:

$$
_{k+1} E_m^{F,\phi} = \frac{1}{2}\left(g_k + A_k\right), \qquad _{k+1} E_m^{F,\psi} = \frac{1}{2}\left(g_k - A_k\right)
\tag{3.15}
$$

Similarly, the hard boundary condition on the backward wave at the nth cell of the domain (Fig. 3.6), at $z = L$ is applied via the expressions:

$$
\begin{aligned}
B_{k+1} = B_{k-1} - 2s\left(\ _k E_{n-1}^{B,\phi} + \ _k E_{n-1}^{B,\psi}\right) \\
+ 2j\kappa s\,\Delta z \mathbf{Sa}\left(\Delta\beta\Delta z/2\right) e^{-j\Delta\beta(M+1/2)\Delta z}\left(\ _k E_n^{F,\phi} - \ _k E_n^{F,\psi}\right) \\
+ 2j\gamma s\,\Delta z \left\{\frac{1}{2}\left|\ _k E_n^{B,\phi} + \ _k E_n^{B,\psi}\right|^2 + \left|\ _k E_m^{F,\phi} + \ _k E_m^{F,\psi}\right|^2\right\},
\end{aligned}
\tag{3.16}
$$

where $B_k = \sqrt{2}\ _k E_N^B$. Then, the scaling and wavelet terms of the backward field, at the nth MRTD scaling cell, are updated according to the scheme:

$$_k E_m^{B,\phi} =\ _k E_m^{B,\psi} = \frac{B_k}{2}$$

Finally, absorbing boundary conditions for backward and forward waves are imposed at $z = 0$, L, via matched layer absorbers. The absorbers are implemented as in [47], by expanding their quadratically varying conductivities in Haar scaling functions. A maximum conductivity $\sigma_{\text{max}} = 0.1$ S/m is used in all subsequent simulations.

3.3.3 Validation

For validation purposes, results of a nonadaptive Haar MRTD code are compared to FDTD. The parameters of the fiber are normalized, with respect to the length L of the filter, as follows: $\kappa L = 4$, $\Delta\beta L = 12$, $\gamma L = 2/3$. The time step is $\Delta t = 0.003125 n_0 L/(10c)$ and two cases for the FDTD cell size are considered: $\Delta z = 0.001L$ and $\Delta z = 0.002L$. Respectively, scaling cell sizes for MRTD are chosen to be $\Delta z = 0.002L$ and $\Delta z = 0.004L$, as the introduction of one wavelet level refines the resolution by a factor of two, thus allowing for the reduction of the number of cells in half. Matched layers of 500 and 250 cells terminate the FDTD and MRTD meshes respectively, with maximum conductivity $\sigma_0 = 0.1$ S/m. Simulation data for both aforementioned cases are provided in Table 3.1. Execution times for the relevant FDTD and MRTD cases over 12,000 time steps (on a Sun Ultra 80 workstation at 500 MHz) reveal that the nonadaptive MRTD code is slower than FDTD by a factor of 11–16%. The reason for this expected slowdown is that the factor of increase in operations per cell between MRTD and FDTD is greater than the ratio of FDTD to MRTD cells (equal to two). The two methods agree well on the peak of the transmitted intensity (which corresponds to the forward wave intensity at the end of the fiber). The slight difference can be attributed to the fact that in the MRTD absorber, the conductivity was assumed constant within each scaling cell and therefore it varied less smoothly than the corresponding FDTD

TABLE 3.1: Validation Data for Nonadaptive MRTD Code

PARAMETER	FDTD, I	MRTD, I	FDTD, II	MRTD, II
Cells	1000	500	500	250
CPU time (s)	43.25	48.23	25.55	29.56
$\max |E_F(z = L)|^2$	10.6042	10.6635	10.3541	10.4283

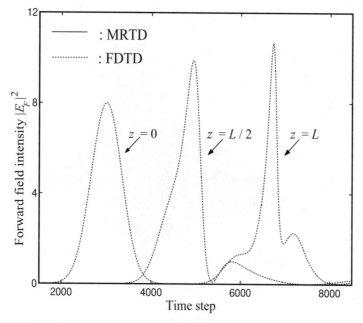

FIGURE 3.7: MRTD and FDTD results for the forward field intensity at the beginning, the middle, and the end of the optical fiber filter (1000 grid points)

absorber conductivity. Moreover, Fig. **??** depicts the pulse compression along the fiber, since it includes pulse waveforms (extracted via 1000 cell FDTD and 500 cell MRTD), as probed at the beginning, the middle and the end of the fiber filter. Again, excellent correlation between the two methods (FDTD and MRTD) is demonstrated. In consequence, the nonadaptive MRTD results can be henceforth used as a measure for accuracy estimation of adaptive MRTD algorithms.

Finally, Figs. **??** and **??** show snapshots of the forward field intensity and the magnitude squared of the forward field wavelet coefficients plotted in space–time coordinates. Evidently, the evolution of both the wavelet coefficients and the field itself takes place along the characteristic line of forward field propagation, $z = ct$. A similar pattern for wavelet behavior was obtained in [15], where nonuniform multiconductor transmission line equations were solved via a biorthogonal wavelet basis. In the following, this observation is utilized for the development of a computationally efficient approach to the problem of thresholding of wavelet coefficients.

3.3.4 Implementation and Performance of Adaptive MRTD

Regarding the development of adaptive algorithms, two questions are addressed in this work: First, the application of thresholding with as few operations (namely checks on wavelet

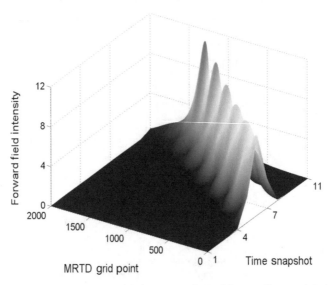

FIGURE 3.8: Spatiotemporal propagation and compression of the nonlinear optical pulse (forward field intensity is shown)

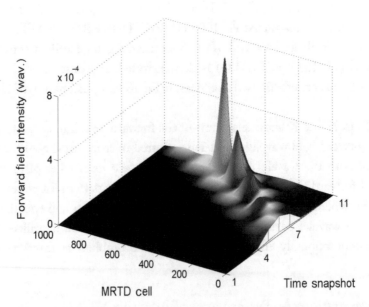

FIGURE 3.9: Spatiotemporal evolution of forward field wavelet coefficients (square of magnitude of wavelet coefficients is shown)

coefficients being absolutely above or below a threshold ϵ) as possible and second, the stable exploitation of the compressed representation of the solution for the reduction in update operations at each time step. For this purpose, the following method is adopted [12, 15, 32], for both forward and backward propagating wave arrays:

- All cells are initialized as being above threshold (referred to as *active*).

- Scaling coefficients are updated throughout the mesh, along with active wavelet terms. In all operations, only scaling and active wavelet terms are taken into account.

- At each time step, the magnitudes of wavelet terms at cells that are designated as active are compared to an absolute threshold ϵ. The corresponding cells remain active only if their wavelets are above the threshold.

- The region of the active cells is extended to all nearest neighbors of the latter. Since only one wavelet level is used here, this implies cells that are immediately to the left or to the right of cells on the border of active regions.

Hence, thresholding checks and update operations are limited to a subset of the field coefficients. Note that the use of the Haar basis and a single wavelet level scheme, keeps the implementation of this adaptive algorithm relatively simple and readily expandable to three dimensions. On the other hand, the complexity of numerical solvers based on higher-order basis functions and multiple wavelet levels has regularly undermined the potential of adaptivity to yield execution times better than FDTD. Furthermore, the frequency of thresholding checks is implicitly dependent on the CFL number s of the simulation, which effectively determines the maximum number of cells that a front may move through in a single time step. Approximately, a thresholding check window should be at most equal to $1/s$, with the previously described procedure strictly corresponding to the value of $s = 1$. Physically, the thresholding algorithm, that was first introduced for the numerical solution of shock wave problems in [32], assumes the evolution of wavelet coefficients along wavefronts defined from the characteristics of a given problem. Then, the purpose of adding pivot elements, to extend the domain of active coefficients, is actually the tracking of these wavefronts as they move throughout the computational domain. The concept of this algorithm is schematically explained in Fig. ??. As shown, adding the pivot elements at the two edges of a wavefront, ensures the tracking of the direction of its movement. Indeed, if that moves to the left, the left pivot element is activated at the "next" time step. Then, the active wavelet region moves to the left, with the wavefront. Otherwise, it follows a movement of the front to the right.

To this end, the second case that was presented in Section ?? is repeated here by using several absolute thresholds and CPU time measurements are carried out. Also, three thresholding

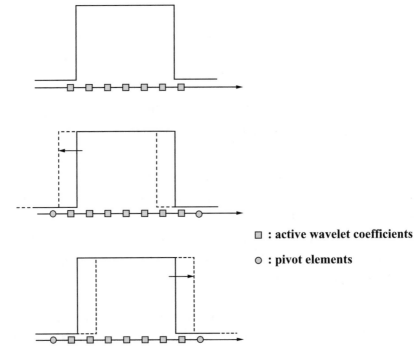

FIGURE 3.10: Graphical presentation of the wavelet-based, front-tracking algorithm

windows are investigated, with the outlined algorithm being applied every one, two, and four time steps respectively. Although the latter was slightly larger than $1/s$, it produced satisfactory (accuracy-wise) results. In all cases, thresholds up to 0.001 led to simulations that clearly resolved the pulse compression and suffered from errors (with respect to the unthresholded MRTD, which is used as a reference solution) of less than 1%.

In Figs. ?? and ??, forward field intensity waveforms sampled at $z = L$, for thresholds 0.1, 10^{-4} and 10^{-7} are shown, along with the incident pulse. The last two are in good agreement both with each other and with the previously presented FDTD and MRTD results, while the first suffers from significant numerical errors, that demonstrate themselves as a ripple corrupting the pattern of the waveform. More explicitly, CPU economy with respect to FDTD and error in the peak transmitted forward field intensity are plotted (with respect to the absolute threshold ϵ), in Figs. ?? and ??. It is thus shown, that the adaptive MRTD code can extract the solution to this problem, at a CPU time reduced (compared to FDTD) by a factor close to 30%, with errors limited at the order of 0.1%.

Thresholding operations in this problem represented a worst-case scenario for the computational overhead that the adaptive algorithm may bring about, for the following reasons: First,

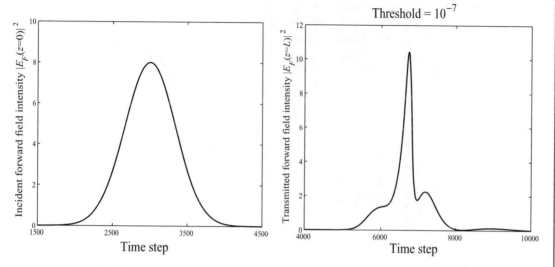

FIGURE 3.11: Incident and transmitted forward field intensity for a threshold of 10^{-7}

the geometry was a one-dimensional one, a significant part of which was almost throughout the simulation occupied by the propagating pulse and second, these operations involved complex numbers and nonlinear terms. Therefore, the fact that an accelerated performance of adaptive MRTD (with respect to FDTD) was achieved is important and demonstrates the potential of the algorithm for larger geometries.

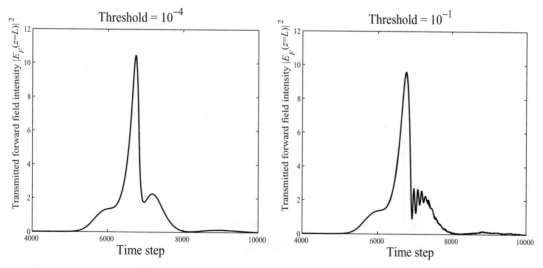

FIGURE 3.12: Transmitted forward field intensity for thresholds of 10^{-4}, 10^{-1}

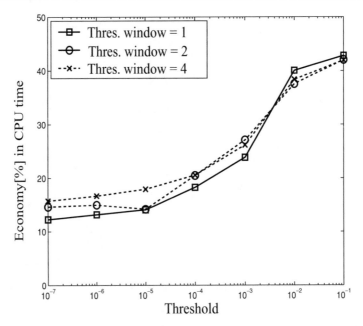

FIGURE 3.13: CPU time economy for the adaptive MRTD code, with respect to FDTD, for absolute thresholds from 10^{-7} to 0.1 applied to Haar MRTD

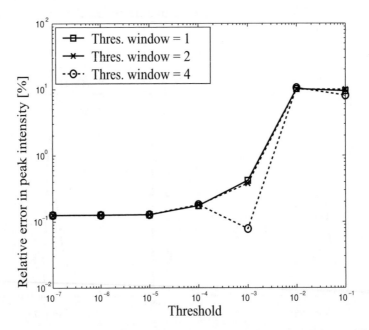

FIGURE 3.14: Relative error (%) in the peak intensity of the transmitted forward field for adaptive Haar MRTD

3.4 CONCLUSIONS

Based on the study of a nonlinear pulse compression by an optical fiber filter, this chapter demonstrated Haar wavelet-based simulations with adaptive meshing that achieved better-than-FDTD execution times. To the extent of the author's knowledge this is the first time when an adaptive, wavelet-based code surpasses the efficiency of the conventional FDTD, not only in terms of memory but also in terms of execution time requirements. The satisfactory performance of the proposed technique stems from its relative simplicity that allows for an efficient implementation of its two components: thresholding tests of wavelet coefficients and operation savings while performing updates of field arrays.

CHAPTER 4

The Dynamically Adaptive Mesh Refinement (AMR)-FDTD Technique: Theory

The finite-difference time-domain (FDTD) method is combined with the adaptive mesh refinement (AMR) technique of computational fluid dynamics to achieve a fast, time-domain solver for Maxwell's equations, based on a three-dimensional, adaptively refinable moving cartesian mesh. This combination allows the resulting technique to adapt to the problem at hand, optimally distributing computational resources in a given domain as needed, by recursively refining a coarse grid in regions of large gradient of electromagnetic field energy. Thus, instead of setting up a statically adaptive mesh, which is the subject of extensive previous research, the idea of a dynamically adaptive mesh is pursued, because of the optimal suitability of the latter to the nature of time-domain electromagnetic simulations. However, the algorithm is still implemented within the context of FDTD, with second-order accurate update equations and is referred to as dynamically AMR-FDTD, to emphasize the dynamic evolution of the underlying mesh. This chapter is aimed at providing the reader with a detailed description of the algorithm, filled with all the information needed for its two- and three-dimensional implementation. The chapter is concluded with a comparison between the dynamically AMR-FDTD and the Haar MRTD approach that was discussed in the last two chapters.

4.1   INTRODUCTION

The finite-difference time-domain technique [6] has been extensively employed in the modeling of microwave integrated circuits [48]. It is especially suitable for wide-band applications since it allows for the characterization of a given structure in a broad frequency range, through a single simulation. However, the FDTD stability and dispersion properties impose severe limitations on the choice of the cell size and the time step of the method, rendering its application to complex structures computationally expensive.

The challenge of accelerating FDTD simulations for practical geometries has been addressed in the past with a variety of static subgridding techniques [10, 44]. According to those, local mesh refinement is pursued in *a priori* defined regions of a computational domain, as dictated by physical considerations. For example, the presence of metallic edges, or high dielectric permittivity inclusions, would call for a locally dense mesh, embedded in a coarser global one. The use of local mesh refinement typically results in significant computational savings compared to the conventional FDTD, despite the fact that its implementation is associated with additional interpolation and extrapolation operations in both space and time.

However, this approach ignores the dynamic nature of time-domain field simulations. In fact, techniques such as FDTD and TLM essentially register the history of a broadband pulse propagating in a device under test, along with its multiple reflections from parts of the latter. Hence, a sharp edge of a microstrip structure is not continuously illuminated by the pulse excitation; on the contrary, it is so for a (potentially small) fraction of the total simulation time, during which a local mesh refinement around it is needed. Therefore, static mesh refinement, which is widely employed in frequency-domain simulations and has been incorporated in commercial finite-element tools, is only a suboptimal solution to the mesh refinement problem in the framework of time-domain analysis.

More recently, a moving-window FDTD (MW-FDTD) method was proposed for the tracking of the forward propagating wave in the two-dimensional terrain environment of a wireless channel [49], similar to ideas previously proposed in [50, 51]. The single moving window used by the method was characterized by fixed size and velocity and therefore, it could not track reflections (which were absorbed by terminating boundaries of the window). As a result, the MW-FDTD is not well-suited for microwave circuit simulations, where the modeling of phenomena as common as signal reflection and branching would require multiple and potentially rotating windows.

In the context of computational fluid dynamics, the technique of adaptive mesh refinement (AMR) was introduced in [52], for the solution of hyperbolic partial differential equations. The application of AMR is based on the use of a hierarchical mesh, recursively developed through the refinement of a coarse root mesh, which covers the entire computational domain. The regions of the computational domain that need further mesh refinement are detected via error estimates or indicators such as gradients of the quantity to be solved for. There may also be dense mesh regions, where the use of a dense mesh is not necessary after a certain time step. These can then be coarsened, again in a recursive manner. Dense and coarse mesh regions are organized via a clustering algorithm that is accompanied by regular checks (every certain time steps) of the error estimates, which guide the process of migration of a cell from one level of resolution to another. This procedure can be associated with the algorithm of the previous chapter, which used wavelet field expansions in order to track the spatiotemporal evolution of shock-wave and nonlinear optical pulse propagation problems, respectively. However, the generalization of such wavelet-based algorithms to three dimensions presents a significant added complexity, while the implicit relation between the actual field values and the wavelet expansion coefficients renders the application of boundary conditions, which are essential for the connection of nested meshes of different resolutions, computationally burdensome.

Note that this chapter's algorithm is a "multi-level" one, in the sense that it can support multiple levels of mesh resolution, by refining the Yee cells of an underlying root mesh by a predefined factor. In our terminology, a two-level scheme would enclose Yee cells of size $\Delta x \times \Delta y \times \Delta z$ and $\Delta x/N_s \times \Delta y/N_s \times \Delta z/N_s$, while an N-level scheme would enclose Yee cells of size $\Delta x/N_s^i \times \Delta y/N_s^i \times \Delta z/N_s^i$, for $i = 0, 1, \cdots N-1$, where the factor N_s (here taken equal to 2) is the *mesh refinement factor*.

In this chapter, the conventional FDTD is combined with the AMR method, in order to formulate an efficient AMR-FDTD technique of superior performance [53]. Instead of using fixed subgrids, this method uses subgrids that are adaptively defined, according to the evolution of field distributions in space and time. As an example, when a Gaussian pulse propagating along a microstrip line is simulated, the adaptive mesh refinement scheme successfully tracks the movement of the pulse, thereby refining only the region that surrounds the propagating pulse. In this case, the AMR accuracy is comparable to that of a uniformly (throughout the entire computational domain) dense mesh FDTD.

The chapter is organized as follows: The general structure of the AMR-FDTD algorithm is presented in Section 4.2, while Section 4.3 refers to the update procedure of the tree of meshes that the algorithm uses in order to adaptively track the field evolution. Section 4.4 outlines how cells that need further mesh refinement or cells that can be removed from the mesh are detected, clustered, and managed by the AMR-FDTD scheme. Finally, Section 4.5 compares AMR-FDTD to the MRTD technique outlining their interesting similarities along with their distinctive differences.

4.2   AMR-FDTD: OVERVIEW OF THE ALGORITHM

In general, a cartesian FDTD mesh occupies a rectangular region,

$$A = \left\{ (x, y, z), x_{a_1} \leq x \leq x_{a_2}, y_{a_1} \leq y \leq y_{a_2}, z_{a_1} \leq z \leq z_{a_2} \right\},$$

terminated by closed or absorbing boundaries. Inhomogeneous material properties can be readily assumed, by letting the dielectric permittivity and magnetic permeability of the structure be generic functions of the space variables $\epsilon = \epsilon(x, y, z)$, $\mu = \mu(x, y, z)$, respectively. Let us consider the subdivision of the domain A in $N_x \times N_y \times N_z$ Yee cells, indexed by a triplet (i, j, k). Each cell occupies a volume $\Delta x \times \Delta y \times \Delta z$, where $\Delta x, \Delta y, \Delta z$ denote the cell sizes in x-, y-, and z-directions. The mesh of these $N_x \times N_y \times N_z$ cells is the coarse grid of region A, that the algorithm under development will selectively and locally refine.

Let us now assume that there are cells within the domain A whose refinement is necessary, according to certain accuracy criteria (the discussion of the latter is deferred to Section 4.4). These cells are first clustered together and then covered by rectangular subregions B_m, $m = 1, 2, \cdots N$ which belong to A. Throughout the algorithmic development of the AMR-FDTD, it will be ensured that these subregions can share planar boundaries, yet they cannot overlap. This is important in order to preserve the possibility of further refinement of these subregions independently from each other, as required by the evolution of the field solution. Each rectangular region B_n is subdivided in Yee cells of dimensions: $\Delta x/2$, $\Delta y/2$, $\Delta z/2$. Hence, a *refinement factor* of 2 is used in every direction, reducing the Yee cell volume of the initial mesh by a factor of 8.

To summarize, a coarse mesh has been defined in the rectangular region A, enclosing finer meshes in rectangular subregions B_n of the latter. The mesh of region A, henceforth referred to as mesh A, will be called the root mesh, or level-1 mesh. The meshes of regions B_m, or meshes B_m, will be called child meshes of A, or level-2 meshes. Recursively, each B_n can be further refined to have its own child meshes, again refining the cell sizes involved by a factor of 2. The hierarchically defined meshes, that the computational domain consists of in an AMR-FDTD simulation, can be assembled in the data structure of a "mesh tree." The mesh tree is regenerated every N_{AMR} time steps through a recursive mesh refinement procedure. An example is shown in Fig. 4.1. There is only one level-1 mesh, which is also called the root mesh and covers the entire computational domain. Each level-$m+1$ mesh is created by refining a subset of the cells of a level-m mesh in a rectangular region by a factor N_s (in each direction). Thus, the two meshes form a child–parent relation. The child meshes of the same parent may share an edge, but they may not overlap otherwise. In Fig. 4.1(b) a solid line corresponds to a child–parent relation, while a dashed line corresponds to a boundary shared by meshes of the same level.

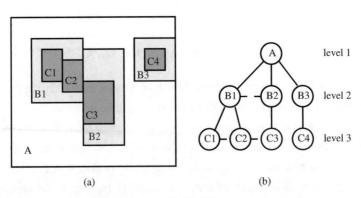

(a)   (b)

FIGURE 4.1: (a) Geometry and (b) data structure of a three-level mesh tree

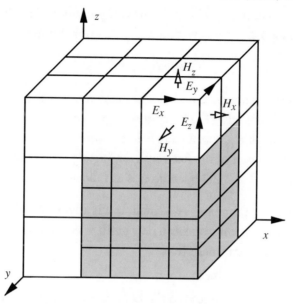

FIGURE 4.2: Sampling points of electric and magnetic field components in a Yee cell. Solid arrows are for electric field, whereas hollow arrows are for magnetic field. The shaded areas represent refined Yee cells

According to the convention of the Yee cell in FDTD (Fig.4.2), electric field components are sampled at the center of the edges of each cell, while magnetic field components are sampled at face centers. The sampling points of a parent mesh may coincide or not with the sampling points of its child mesh, depending on whether the refinement factor is odd or even, respectively. In this work, the refinement factor is 2 (or powers of 2, with respect to the root mesh), hence the grid points of child and parent meshes do not coincide. An alternative case, where the choice of a refinement factor of three renders the parent mesh sampling points also child mesh sampling points can be found in [44].

The main difference between standard subgridded FDTD methods and AMR-FDTD is the dynamic mesh generation which is pursued in the latter, every N_{AMR} time steps. To define the time step of AMR-FDTD, the following observations need to be made. For a level-M mesh, Yee cell dimensions are: $\Delta x/2^{M-1}$, $\Delta y/2^{M-1}$, $\Delta z/2^{M-1}$, where Δx, Δy, Δz are the root mesh Yee cell dimensions. Furthermore, the Courant number is fixed to a constant value s in all meshes. This implies that the root mesh time step Δt (from now on referred to, as AMR-FDTD time step) is given as:

$$\Delta t = s \frac{1}{u_p^{max}} \frac{1}{\sqrt{\dfrac{1}{\Delta x^2} + \dfrac{1}{\Delta y^2} + \dfrac{1}{\Delta z^2}}}, \qquad (4.1)$$

where u_p^{max} is the maximum phase velocity in the computational domain. Applying (4.1) for a mesh of level M, keeping s fixed, yields a time step Δt_M for this mesh, equal to:

$$\Delta t_M = \Delta t / 2^{M-1}. \tag{4.2}$$

For example, level-2 meshes are updated twice as many times as the root mesh. Thus, another shortcoming of the conventional FDTD is addressed; the minimum time step of the algorithm is only used for the update of regions of large field variations, as opposed to the whole domain, a salient feature that is also part of the fixed subgridding algorithms of [10, 44].

The loop of the AMR-FDTD operations is as follows:

1. Check the number of time steps executed. If it is an integer multiple of N_{AMR}, perform adaptive mesh refinement to create a new mesh tree, and carry the field values from the old mesh tree to the new mesh tree.

2. Update fields of the root mesh.

3. Copy fields from the root mesh to the boundary of the child meshes. Update fields of the child meshes 2^{M-1} times, where M is the resolution level of the mesh. Copy fields from child meshes back to the root mesh, for the time steps of the latter.

4. Check whether the maximum time step has been reached. If so, terminate the simulation, otherwise return to step 1.

The next two sections are aimed at explaining these steps in detail.

4.3   MESH TREE AND FIELD UPDATE PROCEDURE IN AMR-FDTD

AMR-FDTD applies the well-known field update equations of FDTD for each mesh, yet the interconnection of the different resolution meshes that march in time at different time steps is an issue to be addressed explicitly. In this section, the types of interfaces that occur in an AMR-FDTD domain are presented, along with their treatment in the update process.

4.3.1   Categorization of Boundaries of Child Meshes

The different categories of child mesh boundaries that may practically occur are shown in Fig. 4.3. To facilitate the presentation of the different cases, a two-dimensional case is discussed (readily extensible to three dimensions). Consider two child meshes B_1 and B_2, embedded in a root mesh A. Three separate cases of boundaries can be identified:

1. Segment $e\,a$: It is defined here as a physical boundary (PB), including absorbing and/or perfect electric conductor boundary conditions.

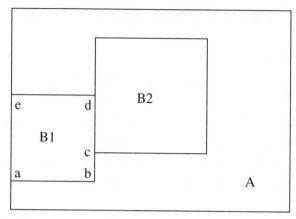

FIGURE 4.3: Types of boundaries of a child mesh

2. Segment *cd*: This is a boundary between "sibling" meshes (SB).

3. Segments *ab, bc, de*: These are boundaries between child and parent meshes (CPB).

Evidently, it is possible that a boundary may belong to more than one of the aforementioned categories. Then, its classification is based on the following hierarchy:

$$PB > SB > CPB.$$

4.3.2 Update of the Mesh Tree

In the following, the update of a two-level mesh tree (with root and level-1 meshes) is discussed. The steps outlined here can be recursively extended to mesh trees of more levels, by considering, for example, level-2 meshes as roots for level 3-meshes and so on. As before, Δt is the time step of the root mesh, while the child level-1 mesh uses a time step $\Delta t/2$. Let us consider the order of updates, assuming that the time-marching procedure has reached the time $t = n\Delta t$. As a result, it is assumed that the root mesh contains the updated values of the electric field component grid points at $n\Delta t$ and those of the magnetic field component grid points at $(n - 1/2)\Delta t$. In addition, the child meshes contain the updated electric and magnetic field values at $n\Delta t$ and $(n - 1/4)\Delta t$, respectively. Then, the following procedure is applied.

1. Backup magnetic field components of the root mesh at $(n - 1/2)\Delta t$. Obtain their values at $(n + 1/2)\Delta t$, by applying the FDTD update equations.

2. Backup the electric field components of the root mesh at $n\Delta t$. Obtain their values at $(n + 1)\Delta t$, by applying the FDTD update equations.

3. For each child mesh, apply the update equations to obtain the magnetic field values at $(n + 1/4)\,\Delta t$.

4. For each child mesh:
   (a) Update the interior (nonboundary) electric field grid points, to obtain their values at $(n + 1/2)\,\Delta t$.
   (b) Update the boundary electric field grid points, at boundaries of the PB, SB, CPB-type, to obtain their values at $(n + 1/2)\,\Delta t$. Obviously, these updates are nontrivial, since they invoke grid points of the root mesh, calculated at time steps of the child mesh. Therefore, interpolation needs to be carried out, in a way that is analyzed in the next subsection.

5. For each child mesh, backup the magnetic field components at $(n + 1/4)\,\Delta t$, and obtain their values at $(n + 3/4)\,\Delta t$, by applying the FDTD update equations.

6. Repeat steps 4a, 4b, to advance the electric field values of the child meshes to $(t + \Delta t)$.

7. For each child mesh:
   (a) Put the spatially interpolated electric field nodal values at $(n + 1)\,\Delta t$ back to the parent mesh, excluding CPB grid points.
   (b) Put the spatially interpolated magnetic field nodal values at $(n + 1/2)\,\Delta t$ back to the parent mesh.

This process can be readily extended to the multilevel case, where meshes of levels $1, 2, 3, \ldots, N_L$ are supported. If N_s is the mesh refinement factor, assuming that the level-1 (root) mesh uses a time step Δt, the level-L meshes use a time step $\Delta t/N_s^{(L-1)}$. For level-L meshes to advance one time step of theirs, level-$(L + 1)$ meshes need to advance N_s of their own time steps to be synchronized with the former. It should be noted that all meshes of the same level need to be updated for a certain time step before advancing to the next time step, because their values are invoked in the update equations of neighboring meshes. The following operations are recursively performed:

1. For each mesh of level L, the magnetic field components are updated.

2. For each mesh of level L, the electric field components are updated.

3. If $L + 1 \leq N_L$, for each mesh of level L, for a time step counter k varying from 0 to $N_s - 1$, the previous two sets of updates are performed.

4. For all the meshes except the root mesh, one can define a "local" time step variable S, which enumerates the time steps executed within one time step of its parent mesh. Hence, this counter will take on values from 0 to $N_s - 1$.

(a) If $S = \lfloor N_s /2 \rfloor$, the magnetic field components of each mesh of level L, are transferred to its parent mesh. Since the node points of parent and child meshes are not necessarily collocated (they are evidently offset if $N_s = 2$), the former are interpolated from the latter.

(b) If $S = N_s - 1$, the electric field components of each mesh of level L are transferred to its parent mesh.

The aforementioned steps can be implemented in a function *update Level*(L, S), which is recursively called to update all meshes with level greater than or equal to L and a local time step S. The main program just needs to call *update Level*$(1, 0)$ for the update of the entire mesh tree for one time step.

4.3.3   Field Updates on CPB-Type Boundaries

The electric field components of a child mesh tangential to its CPB-type boundary (steps 4a, 4b) are obtained from its parent through trilinear interpolation in space and time, as shown in Fig. 4.4. Since such a boundary is characterized by fixing one spatial variable and letting the other two vary, along with time, trilinear interpolation provides the expression employed in all these updates. To interpolate a function $f(\eta, \xi, \tau)$ in the range $0 \leq \eta, \xi < 1, 0 \leq \tau < 1$, by using values of the function at points $(0, 0, 0)$, $(0, 0, 1)$, $(0, 1, 0)$, $(0, 1, 1)$, $(1, 0, 0)$, $(1, 0, 1)$,

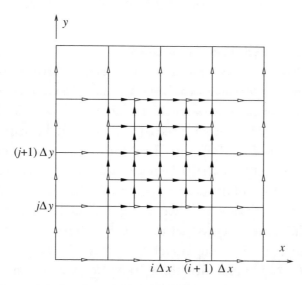

FIGURE 4.4: Sampling points for electric field on CPB-type boundary. The solid arrows are for the child mesh. The hollow arrows are for the parent mesh

$(1, 1, 0)$, and $(1, 1, 1)$, the following formula can be applied:

$$
\begin{aligned}
f(\eta, \xi, \tau) = {} & (1 - \eta)(1 - \xi)(1 - \tau) f(0, 0, 0) \\
& + \eta (1 - \xi)(1 - \tau) f(1, 0, 0) \\
& + \xi (1 - \eta)(1 - \tau) f(0, 1, 0) \\
& + \eta \xi (1 - \tau) f(1, 1, 0) \\
& + (1 - \eta)(1 - \xi) \tau f(0, 0, 1) \\
& + \eta (1 - \xi) \tau f(1, 0, 1) \\
& + \xi (1 - \eta) \tau f(0, 1, 1) \\
& + \xi \eta \tau f(1, 1, 1).
\end{aligned}
\tag{4.3}
$$

As an example, consider the E_x field on a CPB-type boundary located at $z = 0$. The parent mesh has $E_x((i + 1/2) \Delta x, j\Delta y, 0, k\Delta t)$, whereas the child mesh needs $E_x(x, y, 0, t)$. If $(i - 1/2)\Delta x \leq x < (i + 1/2)\Delta x, j\Delta y \leq y < (j + 1)\Delta y, k\Delta t \leq t < (k + 1)\Delta t$, with i, j, k being integers. Then, η, ξ, and τ correspond to the following normalized spatial and temporal variables:

$$
\frac{x}{\Delta x} - i - 1/2, \quad \frac{y}{\Delta y} - j, \quad \frac{t}{\Delta t} - k,
$$

which vary within the interval $[0, 1)$. Since the child mesh regions B_n belong to their root mesh region A, the use of (4.3) allows for the determination of any sampling point of the child mesh from sampling points of its root mesh, enabling the transfer of data which is included in step 4b.

4.3.4   Field Updates on SB-Type Boundaries

First, note that boundaries between child meshes of different levels of resolution can be treated as the CPB boundaries that were discussed above. Therefore, the treatment of SB-type boundaries can be limited to boundaries between child meshes of the same level. Since, by Yee cell convention, the electric field is sampled at cell edge centers and the magnetic field is sampled at cell face centers, two meshes of the same level share tangential electric field and normal magnetic field components at the interface between them (see, e.g., Fig. 4.5).

However, an inspection of the required grid points that the Yee's algorithm invokes in the update of the field components indicated in Fig. 4.5, reveals that only the update of the tangential electric field component needs special handling. Referring to Fig. 4.5, which shows two meshes interfaced at $z = 0$, the update of E_x in mesh 1 is based on the retrieval of the values of $H_{y_1}, H_{y_2}, H_{z_1}, H_{z_2}$. In particular, the index of H_{y_1} in mesh 2 should be transparent to mesh 1. For this purpose, the positions of the SB-type boundaries (between child meshes of the same level) are recorded in a table, after each mesh refinement (i.e., at each N_{AMR} time steps).

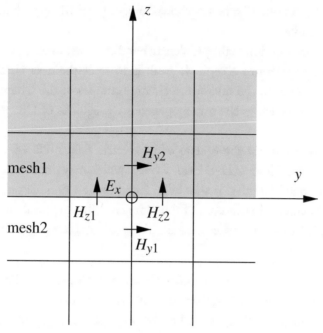

FIGURE 4.5: Interface between two meshes of the same level

4.3.5 Field Updates at Junctions of SB-Type Boundaries

One specific difficulty arising in the update of an SB-type boundary comes from T-junction regions where three sibling meshes share one common edge, and cross-junction regions where four sibling meshes share one common edge. Ideally, an SB-type boundary should be transparent and meshes linked by SB-type boundaries (connecting meshes of the same resolution) should

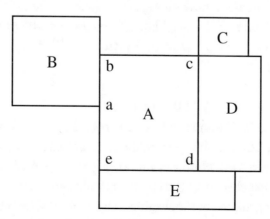

FIGURE 4.6: Typical SB-type boundaries. A, B, C, D, and E are meshes of the same level

behave like one larger mesh. Figure 4.6 shows some typical SB-type boundaries, including junctions a, b, c, d, and e.

A straightforward solution to the problem of handling these junctions, that in dielectrically homogeneous domains operates acceptably well, is to treat them as CPB-type boundaries. This approach is employed in the microwave circuit examples of the Chapter 5. However, for the analysis of structures such as the inhomogeneous waveguides of Chapter 6, the following approach is proposed.

As shown in the figures, the update of an electric field node (E-node) requires four neighboring magnetic field nodes (H-nodes), some of which may not belong to the same mesh. Let us call those external H-nodes. It should be noted that an E-node on a boundary may be shared by 2, 3, or 4 neighboring meshes of the same level. It only needs to be updated in one mesh and then copied to all the other meshes. To avoid double updates, the following three situations are considered:

1. If an E-node on an SB-type boundary is not at a corner of at least one neighboring mesh, it should be updated in that mesh by using one external H-node from the other mesh, then copied to all other meshes. For example, in Fig. 4.6, the E-node a should be updated in A then copied to B, b should be updated in B then copied to A, and d should be updated in E then copied to A and D.

2. If an E-node on an SB-type boundary is at a corner of all the neighboring meshes, the situation can be further divided in two subcases:

   (a) There is at least one pair of meshes forming a cross-junction around the E-node. For example, since meshes A and C form a cross-junction at the E-node c, c can be updated in A by using two external H-nodes from C, or vice versa. Then, c is copied to all the un-updated neighboring meshes.

   (b) There is no pair of meshes forming a cross-junction around the E-node, for example, E-node e. None of the neighboring meshes can obtain two external H-nodes, therefore this E-node has to be treated as a CPB-type boundary and updated accordingly.

4.3.6   Field Updates on PB-Type Boundaries

PB-type boundaries include absorbing and/or PEC boundary conditions. These conditions are enforced in both root and child meshes. For the applications that follow, Mur's first-order boundary condition [54] has been used. However, any other type of boundary condition can be readily incorporated. Perfectly matched layer (PML)-type conditions, would simply extend the computational domain by the number of the absorber cells. Since PMLs are terminated into PECs, the only type of PB boundaries occurring in a PML-terminated domain are those of PECs.

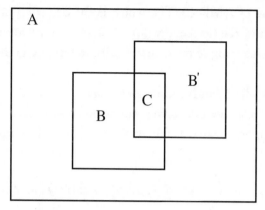

FIGURE 4.7: Transition from old mesh tree to new mesh tree

4.3.7   Transition from Old to New Mesh Tree

Each N_{AMR} time steps of the root mesh, a new mesh tree is created, representing the adaptive mesh regeneration. Field samples stored in the old mesh tree are transferred to the new mesh tree. This is straightforward for the root mesh, since the new tree has the same root mesh as the old tree. On the other hand, the possibility of an overlap between any child mesh of the new tree and any child mesh of the old tree is checked. If there is such an overlap, the field samples in the overlapping region are transferred from the old to the new child mesh. For example, consider the situation shown in Fig. 4.7, where A is the root mesh, B is an old child mesh, B' is a new child mesh. B and B' overlap in region C. Fields of B within C will be copied to B' directly and the rest of B' is initialized by interpolating fields of A. Note that if B' contains boundaries, where conditions, such as Mur's first-order absorbing boundary condition, are applied, then field values of the current *and* the last time step are needed. These are maintained and kept available, according to the proposed algorithm of Section 4.3.2. Again, trilinear interpolating operations are employed to initialize the new mesh.

It is finally noted that source conditions are always enforced in the root mesh. If a child mesh overlaps with the source region, the overlapping part of the source should also be enforced in the child mesh.

4.4   ADAPTIVE MESH REFINEMENT

Up to this point, the features of AMR-FDTD relevant to the enforcement of a nonuniform grid and the implementation of multiple subgrids within a root mesh have been explained. Standard interpolation operations, which are the common characteristic of any subgridding algorithm have been proposed. What distinguishes AMR-FDTD from previous subgridding approaches is the adaptive mesh refinement, which enables the adaptive movement of the subgrids. This section explains the core of AMR-FDTD, which is the detection of cells that need further refinement, the dilation of the detected region, to account for wave propagation during the AMR interval and the clustering of the detected cells, to form child meshes.

4.4.1   Detection of Cells That Need Refinement

Every N_{AMR}-steps, the following calculations are carried out. First, the energy of each (i, j, k) cell of the *root* mesh is approximated as follows (assume that the computation is made at time step n):

$$
W_{i,j,k}^n = \frac{1}{2} \int_{V_{i,j,k}} \left\{ \epsilon \left| \overline{E}(\overline{r}, n\Delta t) \right|^2 + \mu \left| \overline{H}(\overline{r}, n\Delta t) \right|^2 \right\} dv
$$

$$
\approx \frac{1}{2} \left\{ \epsilon_{i,j,k} \left| \overline{E}_{i,j,k}^n \right|^2 + \mu_{i,j,k} \left| \overline{H}_{i,j,k}^n \right|^2 \right\} V_{i,j,k},
$$

(4.4)

where $V_{i,j,k}$ is the volume of cell (i, j, k), $W_{i,j,k}^n$ is the electromagnetic energy in this cell at time step n, and $\overline{E}_{i,j,k}^n$, $\overline{H}_{i,j,k}^n$ are vector electric and magnetic field values at the center of the cell at time step n, which can be approximated by space/time averaging. Then, the gradient of the energy is numerically approximated by a second-order finite-difference expression, as:

$$
\nabla W_{i,j,k}^n = \frac{W_{i+1,j,k}^n - W_{i-1,j,k}^n}{2\Delta x} \hat{x} + \frac{W_{i,j+1,k}^n - W_{i,j-1,k}^n}{2\Delta y} \hat{y}
$$

$$
+ \frac{W_{i,j,k+1}^n - W_{i,j,k-1}^n}{2\Delta z} \hat{z}.
$$

(4.5)

Defining thresholds θ_g and θ_e, a cell (i, j, k) is marked for refinement if both of the following criteria are met:

$$
\left| \nabla W_{i,j,k}^n \right| > \theta_g G^n
$$

$$
W_{i,j,k}^n > \theta_e Q^n,
$$

(4.6)

where

$$
G^n = \max_{i,j,k} \left| \nabla W_{i,j,k}^n \right|
$$

(4.7)

$$
Q^n = \max_{0 \le m \le n} W_{av}^m,
$$

(4.8)

$$
W_{av}^m = \frac{1}{N_x N_y N_z} \sum_{i,j,k} W_{i,j,k}^m.
$$

(4.9)

Therefore, the adaptive mesh refinement is executed in a cell when both an instantaneous and a calculated over the whole simulated time threshold are exceeded. The first criterion takes into account the appearance of energy gradient peaks, while the second ensures that arbitrary field fluctuations, mainly stemming from numerical errors at the final steps of the simulation, will not create an unnecessary refinement process.

On the other hand, as the AMR-FDTD simulation begins, all the electric field components assume zero values, except for the ones excited by the source. The detection of cells to be refined at this time-marching stage is difficult, since the energy gradient is too small to surpass the threshold set. To overcome this difficulty, in addition to the cells detected by thresholding, the cells in the source region are also marked for refinement, for a certain period of time. For example, if a Gaussian excitation of the form:

$$\exp\left(-\left(\frac{t - t_0}{T_s}\right)^2\right)$$

is used (with $t_0 \sim 3T_s$), source region cells are refined up to time $t = 2t_0$.

For the multilevel case, the same algorithm is recursively applied at each mesh level. Assuming again that N_L is the maximum number of levels, the mesh regeneration (and, subsequently, the update of the mesh tree) is pursued as follows:

1. For mesh level L varying from 1 to N_L,
   (a) The maximum average energy Q_L^n and maximum gradient of energy G_L^n for level L is calculated.
   (b) For each mesh of level L, the cells whose energy is greater than $\theta_e Q_L^n$ and gradient of energy is greater than $\theta_g G_L^n$ are marked. If no cells are marked, we return to the previous step. Otherwise, the marked cells are clustered into rectangular regions, and for each rectangle, a level-$(L + 1)$ mesh is created and added to the new mesh tree.

2. For L varying from 2 to the number of levels of the new mesh tree, each level-L mesh is initialized as follows: if it overlaps with any level-L mesh of the old mesh tree, the electric and magnetic field components are obtained from the overlapping region. For the remainder of the mesh (which does not overlap with any other previous child mesh) the electric and magnetic field components are calculated by interpolating their values from the corresponding parent mesh.

3. The previous mesh tree is then removed from memory.

The only mesh that is not subject to regeneration every N_{AMR} time steps is the root mesh.

4.4.2   Modeling of Wave Propagation

The application of the previous criteria may result in a number of cells being marked for refinement. However, this refinement process takes place every N_{AMR} time steps (to avoid loading every time step with the mesh refinement operations). Within these time steps, wave propagation within the computational domain can clearly generate the need for a denser mesh in cells neighboring to the already refined. To capture this field movement, cells within a distance

$$D = \sigma_{AMR}\, N_{AMR}\, c\, \Delta t \qquad\qquad (4.10)$$

from a marked cell are also refined, where c is the speed of light. The factor σ_{AMR} is a predefined, greater-than-one positive real number, called the spreading factor. To facilitate computations, the cell-to-cell distances can be defined in the sense of the Manhattan distance $d(\bar{r}_1, \bar{r}_2) = |x_1 - x_2| + |y_1 - y_2| + |z_1 - z_2|$ instead of the Euclidean distance $d(\bar{r}_1, \bar{r}_2) = (x_1 - x_2)^2 + (y_1 - y_2)^2 + (z_1 - z_2)^2$. Physically, this distance D is the maximum distance that a wave can travel within time equal to $N_{AMR}\, \Delta t$, multiplied by the spreading factor. Note that no assumption is being made as to the direction of the wave velocity, which in general is unknown.

The effect of the spreading factor, as well as the thresholds defined in the previous section will become evident in the numerical results of the next two chapters.

4.4.3   Clustering

Since the cells that are marked for refinement would generally define irregularly shaped regions, they are first covered by a number of boxes, which are then divided in Yee cells. This procedure is called clustering. To evaluate the quality of clustering, the box coverage efficiency is introduced, defined as the ratio of the total volume of the marked cells, covered by a box, to the volume of the box.

For the implementation of this clustering procedure, the methodology proposed in [55] is followed. There are three predefined controlling parameters: the threshold for coverage efficiency θ_c, the minimum dimension of the box D_{min}, and the maximum number of boxes N_{max}. At the beginning, the bounding box enclosing all the marked cells is found and its coverage efficiency is calculated. If the coverage efficiency is greater than θ_c or the dimension is less than D_{min}, the algorithm stops, otherwise the box is split in two boxes across a cut plane. Each of the new boxes is shrunk to just cover the marked cells. Then, the coverage efficiency of each box is calculated and compared with θ_c. Again, either box with coverage efficiency less than θ_c and dimension greater than D_{min} will be split in two. This iterative process continues until the maximum number of boxes is reached.

Figure. 4.8 illustrates the splitting and shrinking of box B. The black dots represent the marked cells. Box A is the root mesh. Box B is split across the cut plane cc' and the resultant two boxes are shrunk to obtain boxes C_1 and C_2. The determination of the position of the cut plane is detailed in [55].

4.4.4   Guidelines for Choosing AMR-FDTD Parameters

As discussed before, the accuracy and efficiency of AMR-FDTD depends on five parameters; θ_e, θ_g, θ_c, σ_{AMR}, and N_{AMR}. Although a more systematic discussion of the dependence of

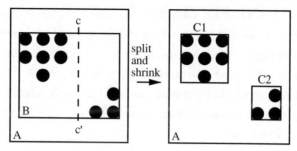

FIGURE 4.8: Splitting and shrinking of boxes (child meshes)

numerical errors of the algorithm on the values of these parameters is included in Section 6.6, some empirical guidelines, qualitatively dictating their choice, can be provided here.

First, it should be noted that the best achievable accuracy by AMR-FDTD is the one of the FDTD scheme applied at a mesh as dense as the finest child mesh in the AMR-FDTD hierarchy of meshes. Therefore, the comments on the accuracy of the technique are meant to be always referred to such an FDTD scheme. Let us also define the AMR-coverage as the ratio of mesh-refined regions to the total volume of the computational domain. In general, decreasing the AMR-coverage will reduce the execution time of the code but will also reduce its accuracy, since the mesh becomes coarser overall.

Furthermore, N_{AMR} determines how frequently the AMR operations are performed. Increasing this parameter leads to less AMR-related operations. A subtle side-effect of a large N_{AMR} is the following: newly generated mesh-refined regions are always extended by a factor proportional to N_{AMR} (see (4.10)), to account for wave propagation between successive mesh refinements. Hence, the coverage increases and thus the execution time. In general, the recommended values of N_{AMR} are between 10 and 50.

The most important source of errors in any static or dynamic mesh refinement scheme is the reflections at a CPB-type boundary. The θ_e parameter directly affects the wave amplitude at such boundaries and as such it is the most important controlling parameter. On the other hand, θ_g ensures that the refined regions enclose large energy gradient variations. In general, $\theta_e = 0.1 - 0.5$, $\theta_g = 0.01 - 0.02$.

Finally, a value of $\sigma_{AMR} = 2$ seems to yield satisfactory results in all cases, while the coverage efficiency threshold θ_c is chosen between 0.6 and 0.8. Note that a large value of the latter enforces the generation of more smaller meshes, improving the AMR-coverage. At the same time, it generates more CPB- and SB-type boundaries, thus increasing the operations related to their management.

These guidelines and the inherent trade-offs in the choice of the parameters are further illustrated in the numerical results of the next two chapters.

4.5   AMR-FDTD AND MRTD: SIMILARITIES AND DIFFERENCES

Essentially, the AMR-FDTD technique belongs to the class of multiresolution time-domain (MRTD) methods, although it does not implicate wavelet basis functions. However, it does implement a multiresolution, space and time adaptive moving mesh in three dimensions, re-generated every certain time steps. However, the MRTD implementations, discussed in this monograph, lack two important features of the AMR-FDTD that render the latter more flexible as a time-domain modeling tool:

1. While the adaptive MRTD operates on nonthresholded field wavelet coefficients, that may be spatially and irregularly distributed, the AMR-FDTD clusters the Yee cells that need mesh refinement and encloses them in rectangular subgrids. This leads to a much more systematic mesh refinement process.

2. In the MRTD implementations that are included in this monograph, as well as in the vast majority of the MRTD literature, a uniform time step is set for the update of all field scaling and wavelet coefficients, regardless of their resolution. On the other hand, AMR-FDTD employs asynchronous updates for the different mesh resolution levels.

However, both AMR-FDTD features can be incorporated in the MRTD technique. An important step toward this end is the work reported in [56]. In general, AMR-FDTD has the advantages of flexibility and versatility and the disadvantages of only second-order accuracy and numerical dispersion that characterize FDTD. In fact, the realistic, two- and three-dimensional applications presented in the next two chapters are aimed at highlighting the former two qualities of the AMR-FDTD. On the other hand, MRTD is a class of numerical techniques that can achieve (with the use of higher-order basis functions, as opposed to the Haar basis) high-order of accuracy, paying the price of complicated enforcement of source and boundary conditions. Hence, despite its more recent introduction the dynamically AMR-FDTD technique is closer to being considered as a mature technique compared to MRTD, which is still related to important and challenging questions when it comes to practical applications.

CHAPTER 5

Dynamically Adaptive Mesh Refinement in FDTD: Microwave Circuit Applications

The dynamically adaptive mesh refinement FDTD technique of the previous chapter is now applied to realistic microwave circuit geometries, namely a microstrip filter, a branch coupler, and a spiral inductor. Through these applications, the salient properties of the technique along with its excellent potential to dramatically reduce FDTD simulation times are shown. Trade-offs between speed and accuracy involved with the application of the algorithm to problems of interest are also discussed.

5.1 INTRODUCTION

In this chapter, a number of microwave circuit analysis examples are presented, aimed at addressing the following two questions that are naturally posed about the dynamically adaptive mesh refinement FDTD algorithm. *First*, what is the effect of the various parameters, defined in the previous chapter on the performance and accuracy of the algorithm and *second*, whether the overhead that AMR-FDTD accumulates from the application of the mesh refinement can still leave some room for computational savings stemming from a reduced overall number of operations. Both questions are negotiated through the application of the technique to three microwave circuit geometries, namely, a microstrip low-pass filter, a branch coupler, and a spiral inductor. In all cases, a maximum of two mesh levels is implemented (or a maximum mesh refinement factor of 8 for a Yee cell of the root mesh). In all these experiments, a two-level AMR-FDTD is implemented, in order to facilitate the presentation of the effect of its parameters on accuracy and execution time. All simulations were executed on an Intel Xeon 3.06 GHz machine.

5.2 MICROSTRIP LOW-PASS FILTER

The first example is a microstrip low-pass filter, shown in Fig. 5.1. The dimensions indicated in the figure are $A = 40$ mm, $B_1 = 2$ mm, $B_2 = 21$ mm, $W = 3$ mm. The substrate thickness is

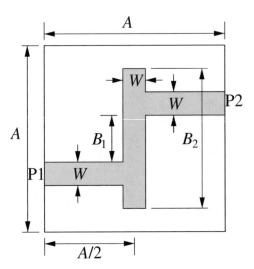

FIGURE 5.1: Microstrip low-pass filter geometry. P1 and P2 denote ports 1 and 2, respectively

0.8 mm and its dielectric constant is 2.2. The dimensions of the air box are 40 mm × 40 mm × 4 mm. Two conventional FDTD simulations were performed for comparison: one using a relatively coarse mesh (40 × 40 × 10 cells), and the second using a dense mesh (80 × 80 × 20 cells). AMR-FDTD simulations use a 40 × 40 × 10 cell root mesh and different controlling parameters for AMR. Both AMR-FDTD and conventional FDTD use Mur's first-order absorbing boundary condition [54]. A voltage source excitation is imposed at 3 mm from the edges. In all simulations, a Courant number of 0.7 is used for determining the time step. While AMR-FDTD and the coarse FDTD technique are run for 4096 time steps, the dense FDTD technique uses 8192 steps, since the time step Δt of the latter is equal to one half the Δt of the former. The AMR parameters are: $\theta_g = 0.02$, $\theta_e = 0.1$, $\theta_c = 0.7$, $N_{AMR} = 10$, $\sigma_{AMR} = 2$.

To demonstrate the evolution of the child meshes over time, the ratio of the volume of mesh-refined areas (occupied by child meshes) to the total volume, referred to as AMR-coverage of the domain and the number of child meshes as a function of time steps are shown in Fig. 5.2. In addition, Figs. 5.3–5.7 show the vertical-to-ground electric field component E_z and child meshes on the plane $z = 0.4$ mm at different time steps. The initial child mesh at $t = 0$ is given in Fig. 5.3. As the wave propagates along the feed line, the coverage of refinement increases, until about $t = 250\Delta t$. Between $250\Delta t$ and $700\Delta t$, the coverage is relatively large (40 to 60%) due to multiple reflections between the two open ends of the microstrip line. As the fields impinge upon the absorbing boundaries of the structure, the field values in the working volume of the domain decrease. Consequently, the spatial field variation becomes smoother, which causes the AMR-coverage to decrease. In fact, after time $t = 1000\Delta t$,

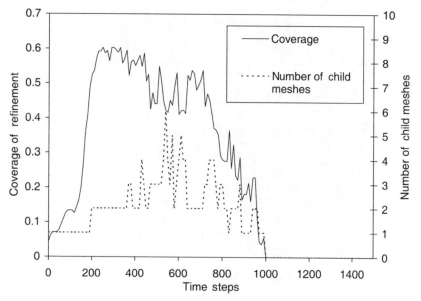

FIGURE 5.2: AMR coverage and number of child meshes at different time steps for the low-pass filter simulation

the coarse root mesh captures the field solution sufficiently well, so that no child meshes are necessary.

Figure 5.8 compares the vertical electrical field at $z = 0.4$ mm and the center of the microstrip line, 3 mm from the right edge, up to 40,000 time steps ($32ns$). The time-series deduced via AMR-FDTD cannot be distinguished from the one of the reference FDTD simulation,

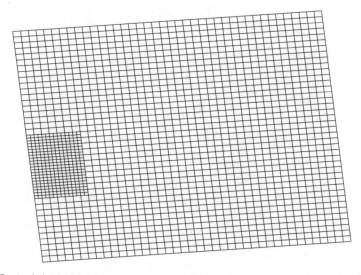

FIGURE 5.3: Vertical electric field magnitude at $z = 0.4$ mm and $t = 0$. One child mesh; AMR coverage is 4.3%

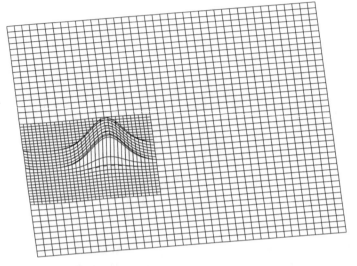

FIGURE 5.4: Vertical electric field magnitude at $z = 0.4$ mm and $t = 100\Delta t$. One child mesh; AMR coverage is 13.4%

whereas the result of FDTD using a coarser mesh has a significant difference. The absence of any late-time instability effects is also noted.

The calculated scattering (S-) parameters and their differences are shown in Fig. 5.9. The plot indicates the excellent approximation provided by AMR-FDTD to the result of FDTD using a dense mesh. This accuracy is quantified in Table **??**, which employs the following

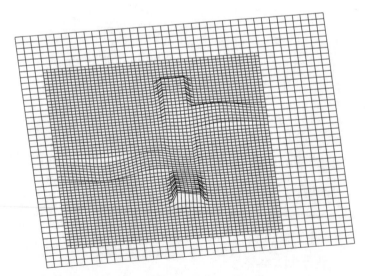

FIGURE 5.5: Vertical electric field magnitude at $z = 0.4$ mm and $t = 200\Delta t$. One child mesh; AMR coverage is 52.5%

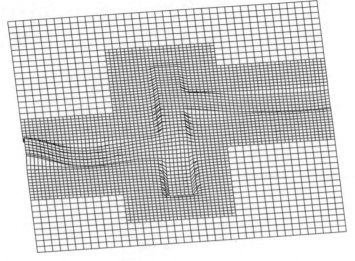

FIGURE 5.6: Vertical electric field magnitude at $z = 0.4$ mm and $t = 500\Delta t$. Three child meshes; AMR coverage is 44.2%

S-parameter error metric:

$$\mathcal{E}_S = \sqrt{\frac{\sum_k \sum_l \sum_m \left| S_{l,m}(f_k) - S_{l,m}^{\text{ref}}(f_k) \right|^2}{\sum_k \sum_l \sum_m \left| S_{l,m}^{\text{ref}}(f_k) \right|^2}}, \qquad (5.1)$$

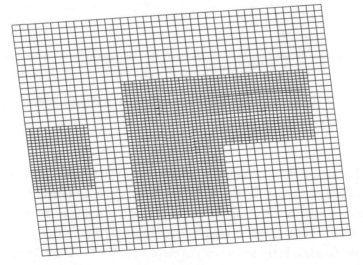

FIGURE 5.7: Vertical electric field magnitude at $z = 0.4$ mm and $t = 800\Delta t$. Three child meshes; AMR coverage is 28.0%

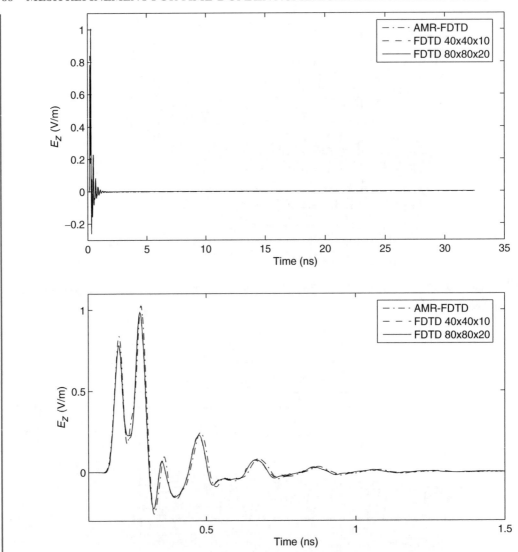

FIGURE 5.8: Vertical electric field magnitude at $z = 0.4$ mm and the center of the microstrip line, 3 mm from the right edge

where, f_k is a discrete frequency within the modeled frequency band (up to 30 GHz), $S_{l,m}$ the (l, m)-element of the scattering matrix of the simulated circuit, as determined by the AMR-FDTD or the coarsely meshed FDTD technique and $S_{l,m}^{ref}$ is the same element, determined by the densely meshed FDTD, which is used as a reference. From Table **??**, it is concluded that AMR-FDTD can closely follow the accuracy of the reference FDTD method, while consuming only 5.4% of its total simulation time. These execution time savings are well above the savings

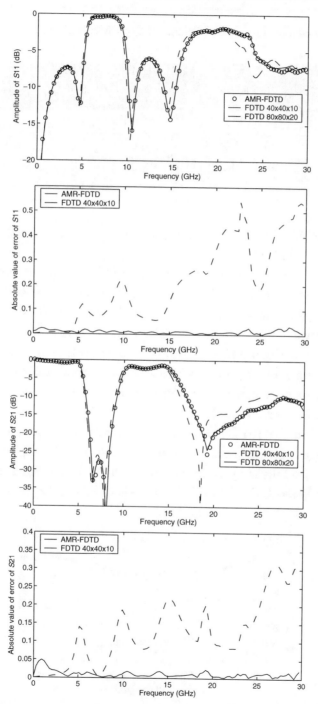

FIGURE 5.9: Comparison of the scattering-parameters of the low-pass-filter of Fig. 5.1, obtained by FDTD and AMR-FDTD

TABLE 5.1: Computation Time and S-Parameter Error Metric $\mathcal{E}_S$ for the Microstrip Low-Pass Filter

METHOD	MESH	TIME STEPS	TOTAL TIME (S)	$\mathcal{E}_s$ (%)
FDTD	$40 \times 40 \times 10$	4096	75	37.3
AMR-FDTD	$40 \times 40 \times 10$	4096	518	2.4
FDTD	$80 \times 80 \times 20$	8192	9669	—

expected from static subgridding algorithms, previously reported in the literature and indicate the potential of AMR-FDTD.

It should be noted that the ratio between the FDTD simulation times for the fine and the coarse mesh is larger than 16:1, which is the ratio between the operations carried out in the two cases. A careful study can show that for small meshes the FDTD simulation times in dense and coarse meshes would tend to follow the 16:1 rule. However, once the mesh size exceeds a certain limit, the simulation time increases faster than 16:1. This is due to the increase in the memory access time for large meshes, which further extends their simulation time. For a thorough investigation of memory-cache related effects in the execution time of time-discrete methods the reader is referred to [57].

The effect of the AMR-FDTD controlling parameters on the simulation time and error of this technique is studied next and results are shown in Figs. ??–??. As expected, the decrease in θ_g and θ_e results in lower errors and longer simulation times. Essentially, as these two thresholds are lowered, the AMR-FDTD tends to become equivalent in operation and

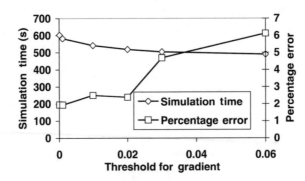

FIGURE 5.10: Effect of threshold for gradient θ_g. The rest of the AMR parameters are: $\theta_e = 0.1$, $\theta_c = 0.7$, $N_{\mathrm{AMR}} = 10$, $\sigma_{\mathrm{AMR}} = 2$

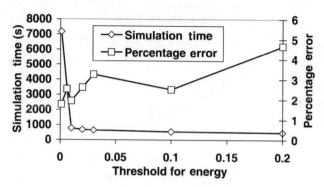

FIGURE 5.11: Effect of threshold for energy θ_e. The rest of the AMR parameters are: $\theta_g = 0.01$, $\theta_c = 0.7$, $N_{AMR} = 10$, $\sigma_{AMR} = 2$

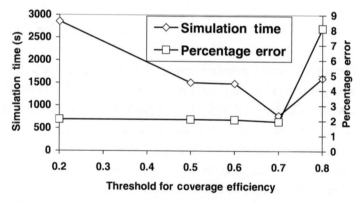

FIGURE 5.12: Effect of threshold for coverage efficiency θ_c. The rest of the AMR parameters are: $\theta_g = 0.01$, $\theta_e = 0.01$, $N_{AMR} = 10$, $\sigma_{AMR} = 2$

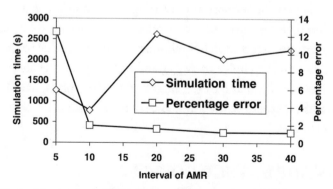

FIGURE 5.13: Effect of interval of AMR N_{AMR}. The rest of the AMR parameters are: $\theta_g = 0.01$, $\theta_e = 0.01$, $\theta_c = 0.7$, $\sigma_{AMR} = 2$

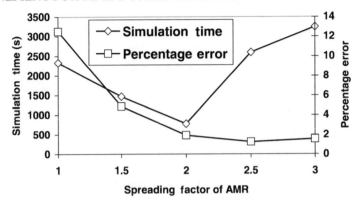

FIGURE 5.14: Effect of spreading factor of AMR σ_{AMR}. The rest of the AMR parameters are: $\theta_g = 0.01$, $\theta_e = 0.01$, $\theta_c = 0.7$, $N_{\text{AMR}} = 10$

performance to the dense FDTD method. On the contrary, increasing these thresholds can reduce the overall computation time, without sacrificing accuracy, up to some point. It should be noted that the percentage error does not decrease monotonically as θ_g and θ_e decrease, because certain child meshes generated by using smaller θ_g or θ_e may have larger reflections due to the irregularity of the shape of the refined regions. This problem can be alleviated by using higher-order interpolation schemes at the mesh boundaries to reduce the reflections [44–58]. It is also noted that there is a sudden increase in the simulation time when θ_e is below 0.01 without any associated improvement in accuracy. At the late stage of the simulation, the field components assume some small values due to the reflections at the CPB- and SB-type boundaries. If θ_e is very small, a large number of child meshes can be generated as a result. Subsequently, the simulation time increases, without any improvement in accuracy. An appropriate choice of θ_e can essentially eliminate this problem. Also, when θ_c reaches 0.8, both the error and the computation time increase. The reason is that the numerical error triggers the automatic generation of multiple spurious child meshes that are clustered independently. As a result, their management by the algorithm adds an overhead that completely eliminates any savings due to the mesh refinement. Trade-off effects related to the choice of N_{AMR} and σ_{AMR} are also evident in Figs. ?? and ??. The appropriate values for N_{AMR} and σ_{AMR}, indicated by these plots, have been used in all the aforementioned numerical experiments. The choice of the AMR parameters is further discussed in this and the next chapter.

5.3   MICROSTRIP BRANCH COUPLER

The microstrip branch coupler geometry of Fig. ?? is analyzed next. The geometric parameters indicated in the figure are as follows: $A = 40$ mm, $w_1 = 2$ mm, $w_2 = 3$ mm, $B_1 = 7$ mm,

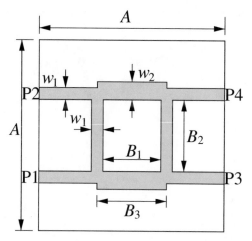

FIGURE 5.15: Microstrip branch coupler geometry. P1, P2, P3, and P4 denote port 1, 2, 3, and 4

$B_2 = 11$ mm, $B_3 = 9$ mm. The thickness of the substrate is 0.8 mm and its dielectric constant is 2.2. The dimensions of the computational domain enclosing the structure are 40 mm × 40 mm × 4 mm. The AMR-FDTD method uses a $40 \times 40 \times 10$ mesh and takes 4096 time steps. A reference FDTD simulation of an $80 \times 80 \times 20$-mesh is used for comparison. The AMR controlling parameters are $\theta_g = 0.01$, $\theta_e = 0.1$, $\theta_c = 0.7$, $N_{AMR} = 10$, $\sigma_{AMR} = 2$. Figures ??–?? show the S-parameters and their errors, as determined by the AMR-FDTD, plain FDTD executed on the AMR root mesh, and the reference FDTD simulation. Comparisons regarding computation times and numerical errors of the AMR-FDTD are included in Table ??. The data demonstrate the capability of AMR-FDTD to deduce the dense FDTD results at a greatly reduced computational cost, which is reflected on an execution time reduction by a factor of 20. Time-domain field waveforms up to 40,000 time steps are shown in Fig. ??. Again, the AMR-FDTD and dense-mesh FDTD waveforms coincide, without any sign of late-time instability.

5.4   MICROSTRIP SPIRAL INDUCTOR

As a last example, the geometry of a spiral inductor of Fig. ?? is analyzed. The parameters of this geometry are: $A_1 = 60$ mm, $A_2 = 40$ mm, $w_1 = w_2 = 2$ mm, $B_1 = 24$ mm, $B_2 = 20$ mm, $B_3 = 18$ mm, $B_4 = 4$ mm. The thickness of the substrate is 0.8 mm and its dielectric constant is 2.2. The dimensions of the computational domain enclosing the structure are 60 mm × 40 mm × 4 mm. The air bridge is 0.8 mm above the substrate. The AMR-FDTD method uses a $60 \times 40 \times 10$ mesh and 8192 time steps. A reference FDTD simulation of a

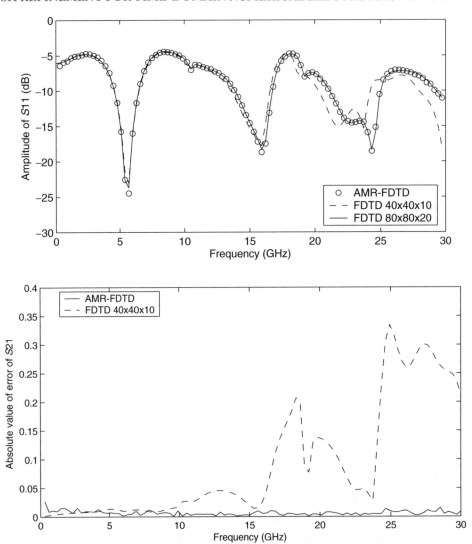

FIGURE 5.16: S_{11} for the microstrip branch coupler geometry of Fig. **??**, as determined by FDTD and AMR-FDTD

$120 \times 80 \times 20$ mesh is used for comparison. The AMR controlling parameters are $\theta_g = 0.001$, $\theta_e = 0.1$, $\theta_c = 0.6$, $N_{AMR} = 50$, $\sigma_{AMR} = 2$. The mesh refinement process is illustrated in Fig. **??**, which shows the effective vertical electric field wavefront tracking achieved by the algorithm. Moreover, Fig. **??** shows the S-parameters and their errors, as determined by the three methods under comparison, while error and execution time data are shown in Table **??**. Time-domain results are also shown in Fig. **??**.

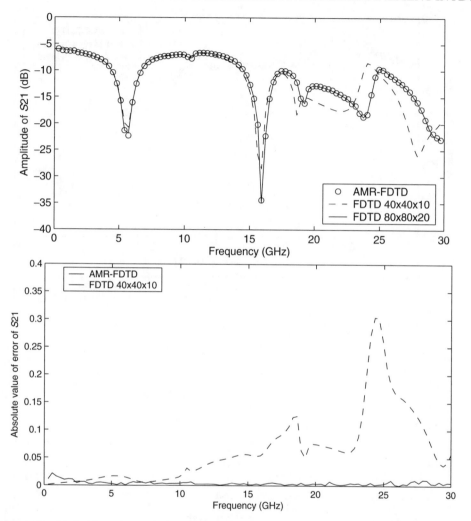

FIGURE 5.17: S_{21} for the microstrip branch coupler geometry of Fig. **??**, as determined by FDTD and AMR-FDTD

Compared to the low-pass filter and the branch coupler, the execution time savings achieved by AMR-FDTD over the conventional FDTD are smaller (yet large, of about 80%), while the associated errors are larger. These effects stem from the highly resonant nature of the spiral inductor, which necessitates the use of a significant number of time steps for the extraction of the S-parameters. At late stages of the simulation, relatively small field values are easily contaminated by spurious reflections at the parent–child mesh interfaces. Note that in this case, the error of the coarse mesh FDTD becomes excessively large as well.

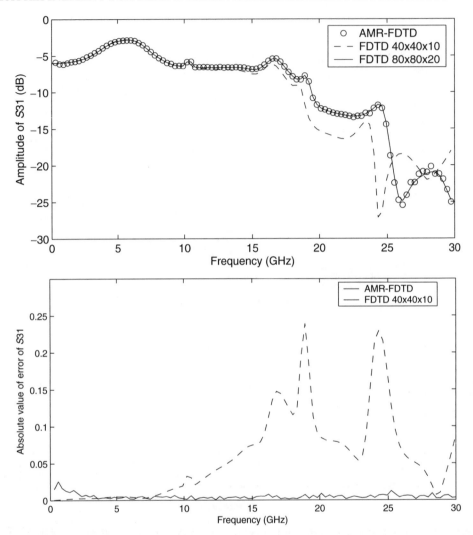

FIGURE 5.18: S_{31} for the microstrip branch coupler geometry of Fig. **??**, as determined by FDTD and AMR-FDTD

5.5   DISCUSSION: STABILITY AND ACCURACY OF AMR-FDTD RESULTS

Based on the previous examples, two further comments are in order. First, the time-domain results accompanying the three numerical experiments (Figs. 5.8, **??**, **??**), demonstrate the absence of late-time instability in the AMR-FDTD. In fact, the convergence of the number of AMR-FDTD child meshes to zero over time, implies that only the root mesh is still present

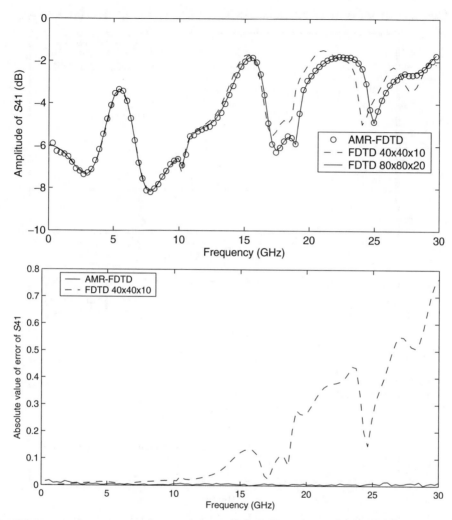

FIGURE 5.19: S_{41} for the microstrip branch coupler geometry of Fig. **??**, as determined by FDTD and AMR-FDTD

TABLE 5.2: Computation Time and S-Parameter Error Metric $\mathcal{E}_S$ for the Microstrip Branch Coupler

METHOD	MESH	TIME STEPS	TOTAL TIME (S)	$\mathcal{E}_s$ (%)
FDTD	$40 \times 40 \times 10$	4096	262	34.6
AMR-FDTD	$40 \times 40 \times 10$	4096	1021	2.43
FDTD	$80 \times 80 \times 20$	8192	19765	—

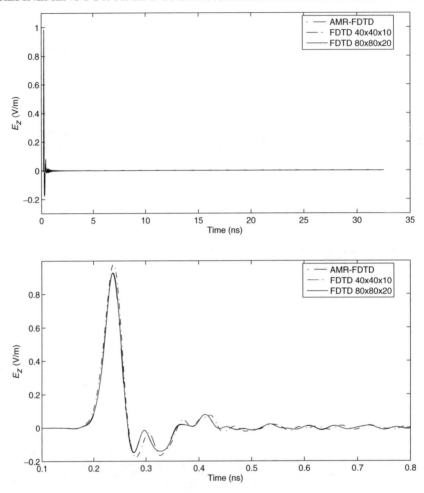

FIGURE 5.20: Vertical electric field magnitude at $z = 0.6$ mm and the center of the microstrip line of port 3, 6 mm from the edge. The excitation is imposed at 3 mm from the edge of port 1

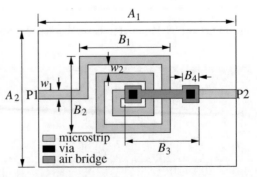

FIGURE 5.21: Spiral inductor geometry. P1 and P2 denote port 1 and port 2, respectively

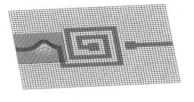

(a) Time-step 100.

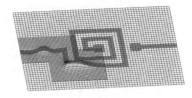

(b) Time-step 200.

(c) Time-step 300.

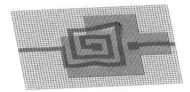

(d) Time-step 400.

FIGURE 5.22: Vertical field evolution and associated mesh refinement in the microstrip spiral inductor, simulated by a two-level dynamic AMR-FDTD

at a late stage of the code. Therefore, no spatial or temporal interpolation operations, which are the primary sources of instabilities in adaptive mesh FDTD codes [44], are applied then. This is an additional advantage of using a dynamically adaptive instead of a statically adaptive mesh in time-domain simulations.

Another aspect of the AMR-FDTD accuracy is associated with the reflections at dense/coarse mesh interfaces. A pulse propagating at a statically refined mesh will be reflected off such an interface, creating numerical errors. Note that in AMR-FDTD, such a pulse would always be enclosed in a dense mesh, while its retro-reflections might encounter dense/coarse grid

TABLE 5.3: Computation Time and S-Parameter Error Metric $\mathcal{E}_S$ for the Microstrip Spiral Inductor

METHOD	MESH	TIME STEPS	TOTAL TIME (S)	$\mathcal{E}_S$ (%)
FDTD	$60 \times 40 \times 10$	8192	411	52.1
AMR-FDTD	$60 \times 40 \times 10$	8192	5276	3.74
FDTD	$120 \times 80 \times 20$	16384	29154	—

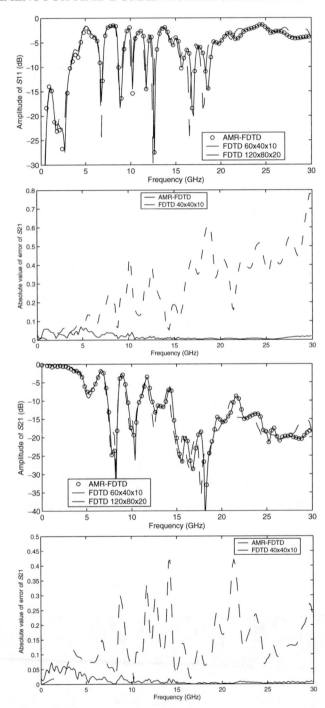

FIGURE 5.23: *S*-parameters for the spiral inductor geometry of Fig. **??**, as determined by FDTD and AMR-FDTD

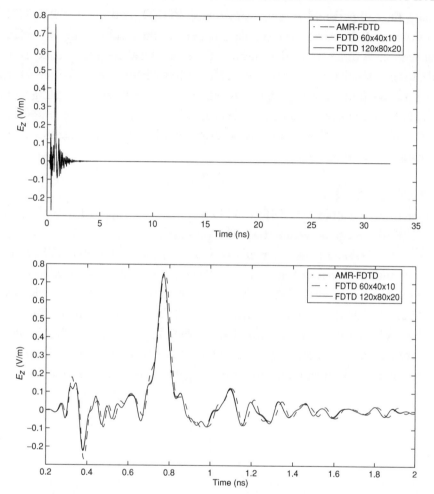

FIGURE 5.24: Vertical electric field magnitude at $z = 0.6$ mm and the center of the microstrip line, 6 mm from the right edge. The excitation is imposed at 3 mm from the left edge of the microstrip line

interfaces before the AMR algorithm creates new meshes for them. This latter case produces errors in AMR-FDTD, which are evidently smaller than those arising in a static subgrid.

However, the application of interpolations in space and time generates a degradation in the FDTD stability factor. This well-known effect, also discussed in [44], can be alleviated by using higher-order interpolation schemes. For the simple trilinear interpolations employed in these computations, the AMR-FDTD stability limit is observed to be 0.9 of the corresponding FDTD one in two-dimensional cases, and 0.85 of the corresponding FDTD one in three-dimensional cases.

Finally, the comparison between coarse/dense FDTD results and AMR-FDTD results in the frequency domain, reveals the standard pattern of the results being in a relatively good agreement with each other up to the middle of the simulated frequency band and diverging afterwards. In these simulations, the coarse grid Yee cell size is about $\lambda_{min}/7.5$ is each dimension. Therefore, up to the frequency $f_{max}/2$, the so-called coarse mesh uses a sampling rate of at least $\lambda(f_{max}/2)/15$, which is dense enough to determine the S-parameters accurately. From that point on, the effect of the FDTD numerical dispersion becomes more severe, leading to the large errors shown in the figures.

5.6   CONCLUSION

In this chapter, the dynamically AMR-FDTD technique was applied to realistic microwave circuit and optical waveguide geometries. The purpose of the applications that were shown, was to demonstrate whether the mesh adaptation overhead of AMR-FDTD can still allow for important computational savings. The conclusion is that it does, because the method is optimally suited to the nature of time-domain simulations. The latter are characterized by spatially and temporally localized phenomena that call for a dense mesh not throughout space and time, but only at the certain time and space they happen. Although microwave-circuit examples were presented, the technique evidently holds a great promise for large-scale optical structure modeling. This direction is further investigated in the next chapter.

CHAPTER 6

Dynamically Adaptive Mesh Refinement in FDTD: Optical Applications and Error Estimates

After presenting microwave circuit applications, the dynamically adaptive mesh refinement FDTD is employed for the modeling of two-dimensional optical waveguide structures, including power splitters, junctions, and ring resonators. The conventional FDTD often requires extremely long simulation times for the characterization of this type of problems; on the contrary, the use of an adaptively moving mesh is shown to result in dramatically accelerated simulations. In all the examples that follow, a multilevel adaptive mesh refinement is pursued. Hence, while in the previous chapter, a maximum to minimum Yee cell edge ratio of two was used, in the following examples this ratio goes up to 64. The use of multiple levels also allows us to compare the efficiency of the mesh refinement algorithm as a function of the number of levels used. The chapter is concluded by extracting error estimates that allow for an educated choice of the AMR parameters.

6.1   MULTILEVEL AMR-FDTD

To demonstrate the application of the AMR-FDTD algorithm with multiple resolution levels, a 2-D TE-mode (with an E_z electric field component only) optical waveguide (shown in Fig. 6.1) is simulated [59]. Its width is 0.3 µm and its dielectric constant is 10.24. The computational domain is 6 µm × 6 µm and a 1 µm thick matched absorber is used to truncate it. Figure 6.2 shows a snapshot of the electric field obtained by AMR-FDTD with a maximum number of levels equal to 4. The excitation is imposed at 1 µm from the left edge of the waveguide and the electric field is recorded at the center of waveguide and 1.5 µm from the right edge. The excitation is a modulated Gaussian pulse of the form:

$$\exp\left(-\left(\frac{y-y_0}{W}\right)^2 - \left(\frac{t-t_0}{T_s}\right)^2\right)\sin\left(2\pi f t\right),$$

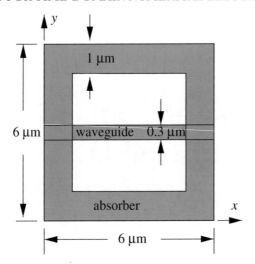

FIGURE 6.1: Optical waveguide geometry

where $T_s = 0.02$ ps, $t_0 = 3T_s$, $f = 200$ THz, $W = 0.7$ μm, $y_0 = 3$ μm. Table 6.1 compares the accuracy and computation time of AMR-FDTD using a root mesh of 120×120 and 2–4 mesh levels, to a reference FDTD simulation using a 1920×1920 mesh and several coarser FDTD schemes. The accuracy is quantified by employing the following time-domain error metric:

$$\mathcal{E}_t = \sqrt{\frac{\sum_k \left| f(t_k) - f^{\text{ref}}(t_k) \right|^2}{\sum_k \left| f^{\text{ref}}(t_k) \right|^2}}, \tag{6.1}$$

where t_k is discrete time within the modeled time range (up to 16 ps), f is the sampled electric field as determined by AMR-FDTD or the coarse mesh FDTD techniques, and f^{ref} is the sampled electric field determined by the reference FDTD simulation. All AMR-FDTD

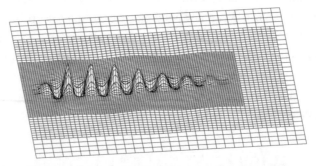

FIGURE 6.2: Vertical electric field magnitude distribution across the optical waveguide of Fig. 6.1, at $t = 100\Delta t$. One level-2 child mesh and one level-3 child mesh are shown

TABLE 6.1: Computation Time and Time-Domain Error Metric $\mathcal{E}_t$ for the Optical Waveguide of Fig. 6.1

METHOD	MESH	NUMBER OF LEVELS	TIME STEPS	TOTAL TIME (S)	$\mathcal{E}_t$ (%)
FDTD	120 × 120	—	2000	6.4	74.9
FDTD	240 × 240	—	4000	66.7	17.2
FDTD	480 × 480	—	8000	635	3.64
FDTD	960 × 960	—	16000	13794	0.60
FDTD	1920 × 1920	—	32000	181997	—
AMR-FDTD	120 × 120	2	2000	31.4	17.0
AMR-FDTD	120 × 120	3	2000	133	3.57
AMR-FDTD	120 × 120	4	2000	641	0.56

simulations use a refinement factor of 2 for successive levels, and $\theta_e = 0.1$, $\theta_g = 0$, $\theta_c = 0.8$, $N_{AMR} = 10$, $\sigma_{AMR} = 2$. Figure 6.2 shows that AMR-FDTD can reach the accuracy of the reference FDTD method in a greatly reduced computation time. The time domain simulation results of the AMR-FDTD with four levels and the reference FDTD code are compared in Fig. 6.3. The AMR-FDTD waveform is not visually discernible from the reference FDTD one.

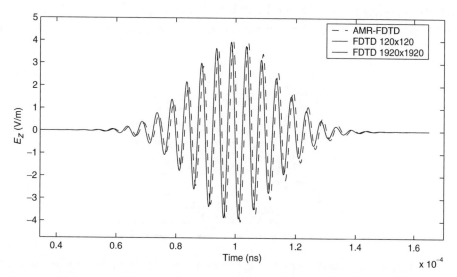

FIGURE 6.3: Temporal waveform of the vertical electric field magnitude at the center of the optical waveguide of Fig. 6.1, 1.5 μm from the right edge

6.2 DIELECTRIC WAVEGUIDE WITH A CORRUGATED PERMITTIVITY PROFILE

A dielectric waveguide with a corrugated permittivity profile, similar to the one presented in [60], is simulated by FDTD and a multilevel dynamic AMR-FDTD [59]. This geometry is a good benchmark application for the AMR-FDTD technique, since the multiple reflections created by the dielectric corrugations enforce strong interactions between spatially distributed meshes, testing the stability and accuracy of the mesh refinement scheme. All interpolations employed are first-order polynomial ones.

The geometry of the waveguide is shown in Fig. 6.4. Its width is 2 μm and the dielectric constant of the host medium is 9. The dielectric constant of the corrugations is 7.9976. The computational domain is 12 μm × 8 μm and is truncated by a 1 μm thick matched absorber. The excitation is imposed 1 μm from the left edge of the waveguide and the electric field is recorded at the center of the waveguide and 1.2 μm from the right edge. The excitation is a modulated pulse of the form:

$$(u(t) - u(t - T)) \sin^2 (\pi t / T) \sin (2\pi f t) e^{-(y - y_0)^2 / W^2}, \qquad (6.2)$$

where $u(t)$ is the unit step function, $f = 193 \, \text{THz}$, $T = 5/f$, $W = 0.7 \, \mu\text{m}$, and $h = 2y_0$ is the height of the waveguide (hence, the pulse is centered in the middle of the guide, along the y-direction). Table 6.2 compares the accuracy and computation time of AMR-FDTD using a root mesh of 240 × 160 and 2–4 mesh levels, to a reference FDTD simulation using a 3840 × 2560 mesh and several coarser FDTD schemes. The simulated time-window in all methods remains constant. Therefore, if one method has a time step Δt and another $\Delta t/2$, the latter will be run for twice the total number of time steps of the former. The accuracy is quantified by employing the time-domain waveform error metric of Eq. (6.1). All AMR-FDTD

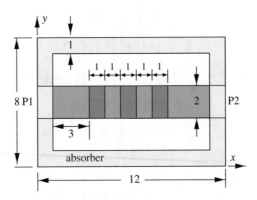

FIGURE 6.4: Geometry of the dielectric waveguide with a corrugated permittivity profile. Dimensions are given in units of μm. P1 and P2 denote ports 1 and 2, respectively

TABLE 6.2: Computation Time and Time-Domain Error Metric $\mathcal{E}_t$ for the Dielectric Waveguide of Fig. 6.4

METHOD	MESH	NUMBER OF LEVELS	TIME STEPS	TOTAL TIME (S)	$\mathcal{E}_t$ (%)
FDTD	240 × 160	—	2000	20.07	157.1
FDTD	480 × 320	—	4000	183.9	41.89
FDTD	960 × 640	—	8000	3544	8.54
FDTD	1920 × 1280	—	16000	50230	0.71
FDTD	3840 × 2560	—	32000	559714	—
AMR-FDTD	240 × 160	2	2000	91.8	41.2
AMR-FDTD	240 × 160	3	2000	246	8.43
AMR-FDTD	240 × 160	4	2000	1253	0.91
AMR-FDTD	240 × 160	5	2000	11337	1.39

simulations use a refinement factor N_s of 2 for successive levels, and $\theta_e = 0.01, \theta_g = 0, \theta_c = 0.8$, $N_{AMR} = 10, \sigma_{AMR} = 2$. Table 6.2 shows that the dynamic AMR-FDTD can reach the accuracy of the reference FDTD method in a greatly reduced computation time. It should be noted that the error no longer decreases once the number of levels of the mesh tree reaches 5, while the execution time also starts increasing then. This trend continues as the number of levels increases and reflects a saturation point beyond which the benefits of the AMR algorithm are compensated for by the errors accumulated through the interpolating operations, as well as the time for their execution. These two problems are interconnected, since the reduction of interpolation errors can be achieved by applying a higher-order polynomial interpolation [44], which in turn increases the operations needed. Therefore, the conclusion of this study is that four levels, combined with first-order interpolations can achieve an impressive reduction in the execution time (by a factor of 446.7), with less than 1% error in the time-domain.

Figure 6.5 compares the time-domain waveforms of AMR and conventional FDTD schemes, while Fig. 6.6 demonstrates the late-time stability of the multilevel dynamic AMR-FDTD. In both figures, the AMR-FDTD uses four levels and a 240 × 160 mesh. The result of Fig. 6.6 can be explained by the following note regarding the stability of the dynamic AMR-FDTD, which is also its fundamental difference from static subgridding schemes. Assume a pulse propagating through a static interface between a coarse and a dense grid. Then, the interpolating operations needed for the update equations would involve large field values. As a

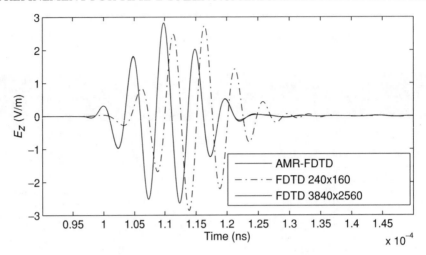

FIGURE 6.5: Electric field at the center of the dielectric waveguide of Fig. 6.4, 1.2 μm from the right edge

result, the associated errors are large, contributing to the well-known late-time instability effects in static subgridding schemes. On the other hand, such an interface would not exist under the dynamic AMR scheme, the reason being that this algorithm *tracks* the propagating wavefronts. Therefore, the pulse of this thought example would be enclosed in a moving dense mesh.

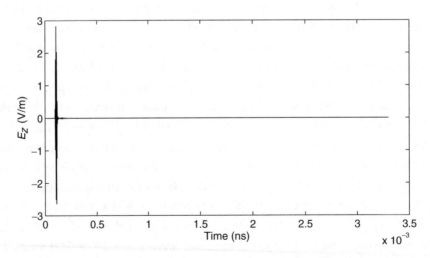

FIGURE 6.6: Electric field at the center of the dielectric waveguide of Fig. 6.4, 1.2 μm from the right edge, obtained with the dynamic AMR-FDTD, for 40,000 time steps, indicating the absence of late-time instability

6.3   DIELECTRIC WAVEGUIDE POWER SPLITTER

To demonstrate the efficiency of the multilevel AMR-FDTD for the simulation of structures with nonconformal dielectric constant distribution, a dielectric waveguide power splitter, shown in Fig. 6.7 [60] is simulated [59]. The dielectric constant of the cross-shaped waveguide is 9, whereas the dielectric constant of the slanted splitter (lens) is 4. It should be noted that due to the slanted orientation of the lens, its dielectric constant profile does not conform to a rectangular mesh. In FDTD simulations using different meshes, the dielectric constant distribution is discretized independently, according to an optimal staircase approximation. The multilevel AMR-FDTD similarly re-discretizes the dielectric constant distribution at different levels. The simulations use the same excitation as (6.2), which is imposed at (1.2 μm, 9 μm), and the electric field is recorded at 1 μm from each port. All AMR-FDTD simulations use a refinement factor of 2 for successive levels, and $\theta_e = 0.01$, $\theta_g = 0$, $\theta_c = 0.8$, $N_{AMR} = 10$, $\sigma_{AMR} = 2$. Table 6.3 shows that AMR-FDTD can reach the accuracy of the reference FDTD method, in a much smaller execution time. In addition, the time-domain results of Figs. 6.8–6.10 demonstrate the late-time stability and accuracy of the method.

Figures 6.11–6.13 show the evolution of the field and the mesh tree, which uses a 60 × 60 root mesh and a maximum number of three levels. The level-1 and level-2 meshes are drawn in the figures, while the regions without grid lines are covered by level-3 meshes (not drawn for visualization purposes). From these figures, the effective wavefront tracking achieved by AMR-FDTD is demonstrated.

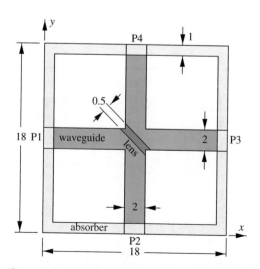

FIGURE 6.7: Geometry of the dielectric waveguide power splitter. Dimensions are given in units of μm. P1, P2, P3, and P4 denote ports 1, 2, 3, and 4, respectively

TABLE 6.3: Computation Time and Time-Domain Error Metric $\mathcal{E}_t$ for the Dielectric Waveguide Power Splitter for Fields Recorded at Port 2

METHOD	MESH	NUMBER OF LEVELS	TIME STEPS	TOTAL TIME (S)	$\mathcal{E}_t$ (%)
FDTD	360 × 360	—	4000	148.92	168
FDTD	720 × 720	—	8000	2313	41.0
FDTD	1440 × 1440	—	16000	39181	2.85
FDTD	2880 × 2880	—	32000	447497	—
AMR-FDTD	360 × 360	2	4000	394.6	41.0
AMR-FDTD	360 × 360	3	4000	816.6	2.98
AMR-FDTD	360 × 360	4	4000	3683	1.07

6.4 DIELECTRIC WAVEGUIDE Y-JUNCTION

A dielectric waveguide Y-junction, shown in Fig. 6.14 is simulated [59]. The dielectric constant of the waveguide is 9 and the dielectric constant of the surrounding medium is 1. The computational domain is 25 μm × 10 μm and a 1 μm thick matched absorber is used to truncate it. The excitation is imposed at 1.2 μm from the edge of port 1 and the electric field is recorded at the center of the waveguide, 2 μm from the edge of port 2. The excitation is a modulated pulse

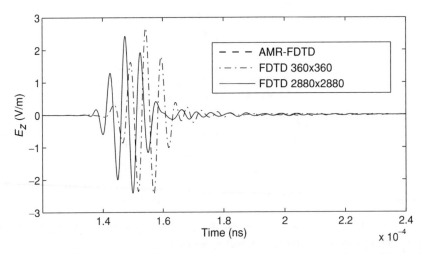

FIGURE 6.8: Electric field at 1 μm from the edge of port 2 of the power splitter of Fig. 6.7

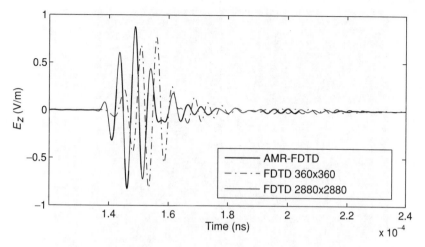

FIGURE 6.9: Electric field at 1 μm from the edge of port 3 of the power splitter of Fig. 6.7

of the form:

$$e^{-(y-y_0)^2/W^2-(t-t_0)^2/T^2} \sin(2\pi f t), \qquad (6.3)$$

where $f = 200$ THz, $T = 0.02$ ps, $t_0 = 3T$, $W = 1$ μm, and y_0 corresponds to the center of the waveguide. Table 6.4 compares the accuracy and computation time of AMR-FDTD using a root mesh of 250×100 and 2–4 mesh levels, to a reference FDTD simulation using a

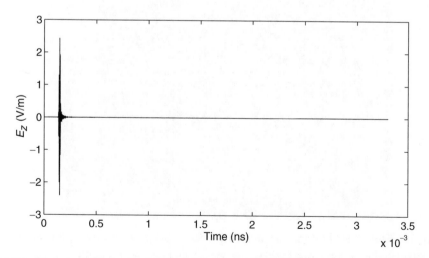

FIGURE 6.10: Electric field obtained with the dynamic AMR-FDTD at 1 μm from the edge of port 2 of the power splitter of Fig. 6.7, obtained with the dynamic AMR-FDTD, for 40,000 time steps, indicating the absence of late-time instability

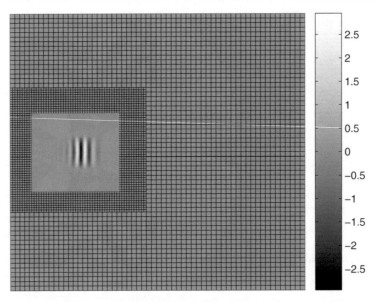

FIGURE 6.11: Electric field magnitude distribution across the power splitter of Fig. 6.7, at $t = 100\Delta t$. There are three meshes in total: one level-1, one level-2, and one level-3 mesh

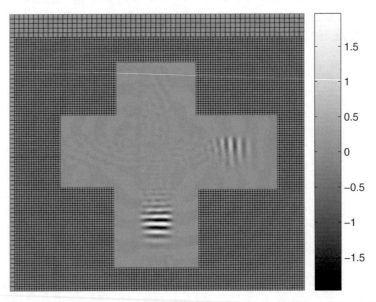

FIGURE 6.12: Electric field magnitude distribution across the power splitter of Fig. 6.7, at $t = 300\Delta t$. There are three meshes in total: one level-1, one level-2, and one level-3 mesh

FIGURE 6.13: Electric field magnitude distribution across the power splitter of Fig. 6.7, at $t = 400\Delta t$. There are eight meshes in total: one level-1, three level-2, and four level-3 meshes (tracking transmitted and reflected wavefronts)

2000×800 mesh and several coarser FDTD schemes. The AMR parameters are: $\theta_e = 0.001$, $\theta_g = 0, \theta_c = 0.8, N_{AMR} = 10, \sigma_{AMR} = 2$. All AMR-FDTD simulations use a refinement factor of 2 for their successive mesh levels. Table 6.4 shows that AMR-FDTD can achieve the accuracy of the reference FDTD method in a dramatically reduced computation time.

Field waveforms in time, extracted by a four-level AMR-FDTD, with a 250×100 root mesh, and the reference FDTD simulation are shown and compared in Fig. 6.15, demonstrating

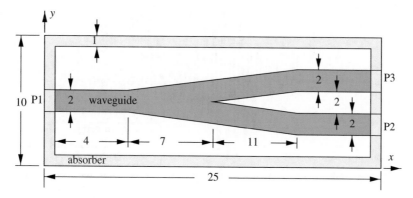

FIGURE 6.14: Geometry of the dielectric waveguide Y-junction. The dielectric constant of the waveguide is 9 and the dielectric constant of the surrounding medium is 1. Dimensions are given in units of μm. P1, P2, and P3 denote ports 1, 2, and 3, respectively

TABLE 6.4: Computation Time and Time-Domain Error Metric $\mathcal{E}_t$ for the Dielectric Waveguide Y-Junction for Fields Recorded at Port 2

METHOD	MESH	NUMBER OF LEVELS	TIME STEPS	TOTAL TIME (S)	$\mathcal{E}_t$ (%)
FDTD	250 × 100	—	2500	15.7	166.7
FDTD	500 × 200	—	5000	155.0	174.4
FDTD	1000 × 400	—	10000	1296	77.8
FDTD	2000 × 800	—	20000	23418	—
AMR-FDTD	250 × 100	2	2500	184.7	174.5
AMR-FDTD	250 × 100	3	2500	646.1	78.4
AMR-FDTD	250 × 100	4	2500	2665	0.55

the excellent agreement of the AMR-FDTD approach to the reference data, despite its reduced numerical cost. A waveform extracted by a conventional FDTD implemented on the root mesh of the AMR-FDTD domain alone is appended, to indicate the accuracy improvement, brought about by the mesh refinement. Moreover, Fig. 6.16 shows that the multilevel AMR-FDTD is free from the late instability problems that characterize static subgridding techniques.

Figures 6.17–6.20 show the evolution of the electric field and the mesh tree, which uses a 125 × 50 root mesh and maximum level of 3. The level-1 mesh is drawn in the figures and

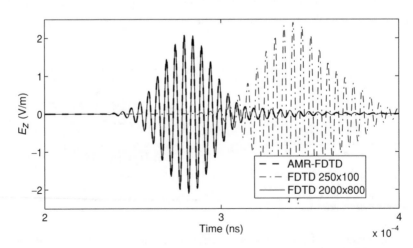

FIGURE 6.15: Electric field at the center of the dielectric waveguide Y-junction of Fig. 6.14, 2 μm from the edge of port 2

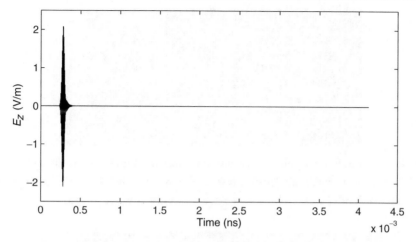

FIGURE 6.16: Electric field at the center of the dielectric waveguide Y-junction of Fig. 6.14, 2 μm from the edge of port 2, obtained with the dynamic AMR-FDTD, for 25,000 time steps, indicating the absence of late-time instability

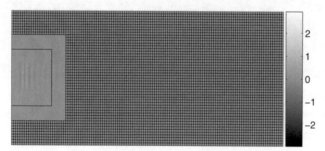

FIGURE 6.17: Electric field magnitude distribution across the Y-junction of Fig. 6.14, at $t = 100\Delta t$. There are three meshes in total: one level-1, one level-2, and one level-3 mesh

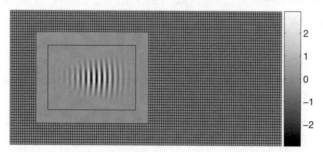

FIGURE 6.18: Electric field magnitude distribution across the Y-junction of Fig. 6.14, at $t = 400\Delta t$. There are three meshes in total: one level-1, one level-2, and one level-3 mesh (enclosed in the black solid line)

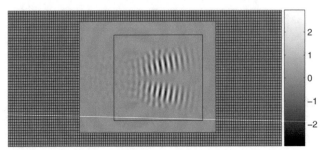

FIGURE 6.19: Electric field magnitude distribution across the Y-junction of Fig. 6.14, at $t = 600\Delta t$. There are three meshes in total: one level-1, one level-2, and one level-3 mesh (enclosed in the black solid line). Note that level-2 and level-3 meshes track the pulse into the junction

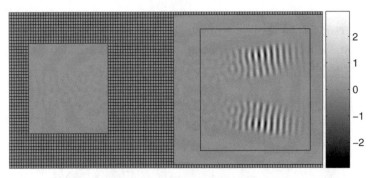

FIGURE 6.20: Electric field magnitude distribution across the Y-junction of Fig. 6.14, at $t = 800\Delta t$. There are four meshes in total: one level-1, two level-2, and one level-3 mesh (enclosed in the black solid line)

the regions without grid lines are covered by level-2 and level-3 meshes, which are not drawn because they are too dense. The rectangular boxes embedded in the solid gray regions indicate the boundaries of level-3 meshes. These figures indicate the wavefront tracking achieved by the dynamic AMR-FDTD.

6.5   DIELECTRIC RING RESONATOR

A highly resonant THz structure of a dielectric ring resonator coupled to two dielectric waveguides, previously modeled in [61] and shown in Fig. 6.21, is simulated [59]. The width of the waveguide and the ring is 0.3 μm, and their dielectric constant is 10.24. The external radius of the ring is 2.5 μm. The edge-to-edge distance between the ring and the waveguide is 0.232 μm. The computational domain is 12 μm × 12 μm and it is truncated by a 1 μm thick matched absorber. The excitation is a modulated pulse of the form:

$$e^{-(y-y_0)^2/W^2-(t-t_0)^2/T^2} \sin(2\pi f t), \tag{6.4}$$

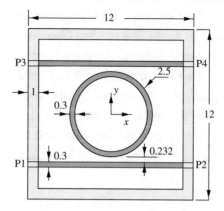

FIGURE 6.21: Geometry of the dielectric ring resonator coupled to two dielectric waveguides. Dimensions are in units of μm. P1, P2, P3, and P4 denote ports 1, 2, 3, and 4, respectively

where $f = 200$ THz, $T = 0.02$ ps, $t_0 = 3T$, $W = 0.7$ μm, and y_0 corresponds to the midsection of the waveguide. The excitation is imposed 1.1 μm from the end of the lower left waveguide port, and the electric field is recorded at 2 μm from the edge of each port. All AMR-FDTD simulations use a refinement factor of 2 for successive levels, and $\theta_e = 0.01$, $\theta_g = 0$, $\theta_c = 0.8$, $N_{AMR} = 10$, $\sigma_{AMR} = 2$. Table 6.5 compares accuracy and execution times of AMR-FDTD and conventional FDTD for this large-scale computational problem. Again, a four-level scheme can reproduce the results of the reference FDTD simulation within a smaller execution time (by a factor of 33), at a relative time-domain error of fields recorded at port 2 of 1.56%.

TABLE 6.5: Computation Time and Time-Domain Error Metric $\mathcal{E}_t$ for the Dielectric Ring Resonator for Fields Recorded at Port 2

METHOD	MESH	NUMBER OF LEVELS	TIME STEPS	TOTAL TIME (S)	$\mathcal{E}_t$ (%)
FDTD	120 × 120	—	20000	60.71	187.0
FDTD	240 × 240	—	40000	703.1	163.6
FDTD	480 × 480	—	80000	6904.4	37.2
FDTD	960 × 960	—	160000	133993	—
AMR-FDTD	120 × 120	2	20000	466.5	163.6
AMR-FDTD	120 × 120	3	20000	1080	36.7
AMR-FDTD	120 × 120	4	20000	4096	1.56

The time-domain results of Figs. 6.22–6.24 demonstrate the late-time stability and accuracy of the method. It should be noted that the first Gaussian pulse arrives at port 3 at about 0.2 ps. Before that, the field at port 3 is due to the disturbance caused by the excitation, which propagates through free space. Since this disturbance is very small compared with the main pulse propagating along the waveguide, it is not tracked by the finest mesh and therefore it induces a relatively large error. However, the pulses circulating the ring resonator and coupled to the waveguides are tracked by the finest mesh and therefore their amplitude and group velocity are very accurately resolved even at very late stages of the simulation.

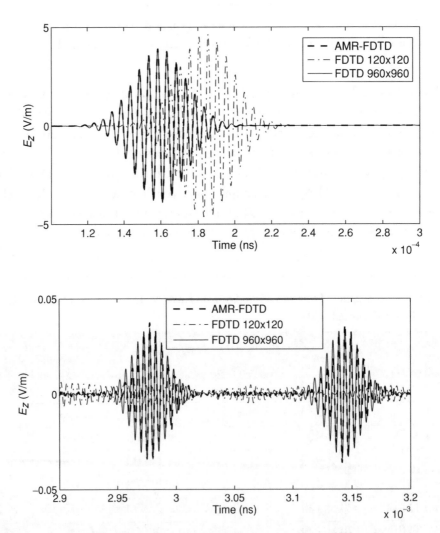

FIGURE 6.22: Electric field at 2 μm from port 2 of the ring resonator of Fig. 6.21

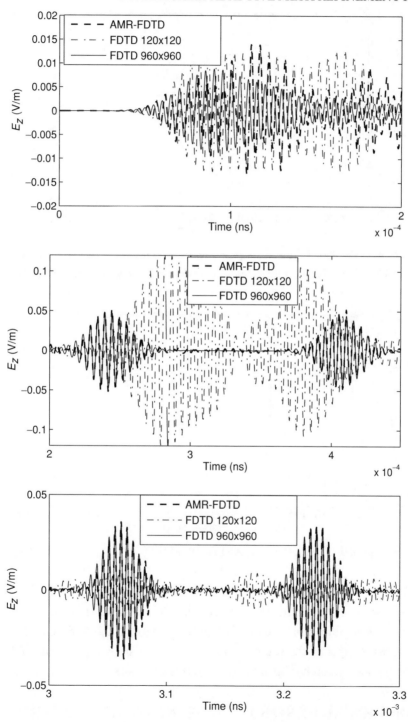

FIGURE 6.23: Electric field at 2 μm from port 2 of the ring resonator of Fig. 6.21

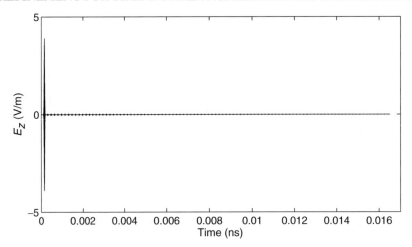

FIGURE 6.24: Electric field at 2 μm from port 2 of the ring resonator of Fig. 6.21, obtained with the dynamic AMR-FDTD, for 100,000 time steps, indicating the absence of late-time instability

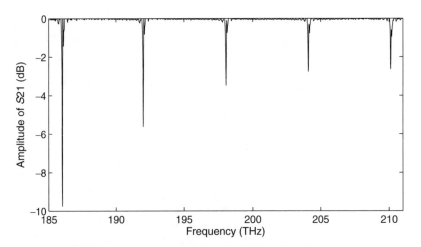

FIGURE 6.25: S_{21} obtained with the four-level AMR-FDTD

Finally, the S_{21} obtained by using AMR-FDTD is shown in Fig. 6.25, which agrees quite well with the results of [61], extracted by FDTD. The AMR-FDTD uses a 120×120 root mesh and four levels. Since the structure is highly resonant, 100,000 time steps are executed. The resonant frequencies of the present method and [61] are compared in Table 6.6, which indicates a very good agreement between the two sets of results.

6.6   NUMERICAL ERROR ESTIMATION AND CONTROL

There are five parameters controlling the accuracy and computation time of AMR-FDTD, as described by the previous sections. To use AMR-FDTD as a CAD technique, a clear set of

TABLE 6.6: Resonant Frequencies (in THz) of the
Dielectric Ring Resonator

ORDER	AMR-FDTD	FDTD [61]
25	186.01	185.85
26	192.03	191.88
27	198.06	197.90
28	204.08	203.92
29	210.11	209.94

guidelines for the choice of the five parameters is needed. The main objective of this section is to produce such guidelines, that would allow a user to determine the AMR-FDTD parameters for a given error tolerance [59].

The numerical error whose dependence on the AMR-FDTD parameters is sought for is defined as follows, for a 2-D TE case with field components (E_z, H_x, H_y). Given a computational domain, an array of probes is considered, where field values are recorded at each time step. The reference solution, that AMR-FDTD is compared to, is provided by applying the FDTD method in a uniform mesh, at a Yee cell resolution equal to the maximum achievable resolution of the AMR-FDTD. This is chosen to be dense enough to guarantee convergence. Then, if a field component recorded at the mth probe at time t_k is denoted by $f_m(t_k)$ and $f_m^{\text{ref}}(t_k)$ the corresponding reference field value, the error is defined as:

$$\mathcal{E} = \sqrt{\frac{\sum\limits_{k,m} \left| f_m(t_k) - f_m^{\text{ref}}(t_k) \right|^2}{\sum\limits_{k,m} \left| f_m^{\text{ref}}(t_k) \right|^2}} \tag{6.5}$$

Note that different probes encounter different field waveforms, but also mesh configurations, due to the dynamic nature of mesh evolution in AMR-FDTD. Therefore, this metric is much more effective in capturing generic error effects, than a norm that would be based on field sampling at a port of a device, or the error in a frequency domain quantity (such as characteristic impedance or propagation constant). This approach is analogous to a Monte-Carlo-type simulation and is aimed at rendering the obtained error less dependent on the device under test. Furthermore, such an approach is the only feasible, given the nature of the problem at hand. Contrary to static subgridding methods, the extraction of analytical error bounds seems to be impossible, due to the arbitrary distribution of sub-meshes in a domain. Finally, the same methodology can be easily extended to 2-D TM cases and 3-D cases.

While the AMR-FDTD error performance on a number of structures was studied, results obtained from simulating three dielectric waveguide ones are shown. These are the power splitter, corrugated permittivity profile waveguide and the Y-junction, presented and simulated in the previous section.

First, the effect of N_{AMR}, σ_{AMR}, and θ_c on the accuracy and simulation time is investigated. Figure 6.26 includes error-simulation time curves, deduced by changing one of these

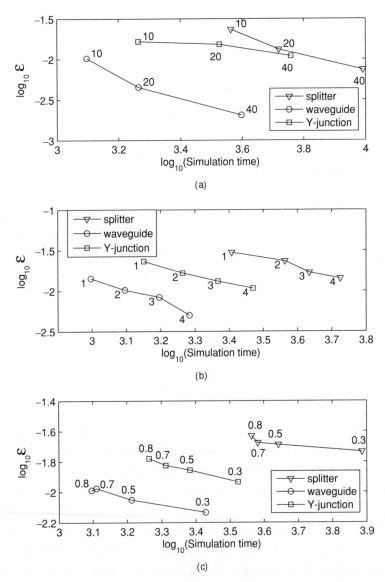

FIGURE 6.26: Error versus simulation time for variable (a) N_{AMR}, (b) σ_{AMR}, (c) θ_c

three parameters and fixing the rest. All figures refer to four-level AMR-FDTD simulations, but two-and three-level simulation results exhibit a similar behavior. The general pattern of these curves suggests that although increasing N_{AMR} and σ_{AMR} or decreasing θ_c may improve accuracy (e.g., as N_{AMR} is increased the AMR-FDTD tends to become equivalent to the reference FDTD technique), this improvement reaches a plateau beyond which, any change in the parameters merely increases the execution time. In a number of versatile situations, the choice of $N_{AMR} = 10$, $\sigma_{AMR} = 2$, and $\theta_c = 0.7$ balances accuracy and efficiency well. This conclusion is also supported by Fig. 6.26. Essentially, these parameters do not explicitly control the mesh refinement procedure, which is why their choice is rather straightforward. This is not the case for θ_e, θ_g that are studied next.

Figure 6.27 shows the dependence of the error $\mathcal{E}$ on θ_e and θ_g for all the cases studied. As expected, the error decreases as θ_e or θ_g decreases. For very small values of θ_g, the error is mainly a function of θ_e and vice versa, since marking a cell for refinement depends on *both* thresholds. Based on this data $\mathcal{E}(\theta_e, \theta_g)$ can be approximated as:

$$\tilde{\mathcal{E}}(\theta_e, \theta_g) = C_1\mathcal{E}(0, \theta_g) \arctan\left(\frac{\theta_g}{\theta_e}\right)^{C_2}$$
$$+ C_3\mathcal{E}(\theta_e, 0) \arctan\left(\frac{\theta_e}{\theta_g}\right)^{C_4}, \tag{6.6}$$

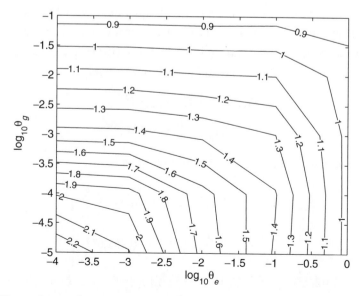

FIGURE 6.27: Dependence of $-log_{10}\mathcal{E}$ on θ_e and θ_g

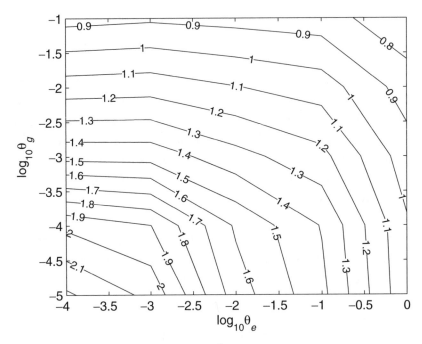

FIGURE 6.28: Curve-fitted dependence of $-log_{10}\tilde{\mathcal{E}}$ on θ_e and θ_g based on (6.6)

where C_1, C_2, C_3, and C_4 are fine-tuned to minimize the difference between $\tilde{\mathcal{E}}$ and $\mathcal{E}$. As an example, Fig. 6.28 shows the curve-fitted error obtained by (6.6), where $C_1 = C_3 = 1.02$, $C_2 = 0.05$, $C_4 = 0.01$, which depicts a very good agreement with Fig. 6.27. Therefore, this approximation, requiring only $\mathcal{E}(\theta_e, 0)$ and $\mathcal{E}(0, \theta_g)$, is useful as sufficiently accurate.

By simulating typical optical waveguide structures, two empirical error bound functions are obtained for the special cases $\theta_e = 0$ and $\theta_g = 0$, respectively,

$$\mathcal{E}_{bg}(\theta_e = 0, \theta_g) = 0.4\theta_g^{0.35}, \tag{6.7}$$

$$\mathcal{E}_{be}(\theta_e, \theta_g = 0) = 0.15\theta_e^{0.37}. \tag{6.8}$$

Following (6.6), we define a general error bound as

$$\mathcal{E}_b(\theta_e, \theta_g) = C_1\mathcal{E}_{bg}(\theta_e = 0, \theta_g) \arctan\left(\frac{\theta_g}{\theta_e}\right)^{C_2} \\ + C_3\mathcal{E}_{be}(\theta_e, \theta_g = 0) \arctan\left(\frac{\theta_e}{\theta_g}\right)^{C_4} \tag{6.9}$$

where $C_1 = C_3 = 1.02$, $C_2 = 0.05$, $C_4 = 0.01$. This general error bound is shown in Fig. 6.29 and can be used to determine θ_e and θ_g given the desired accuracy.

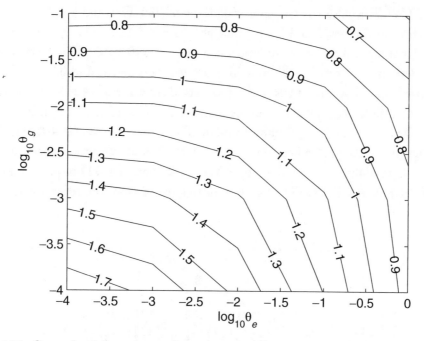

FIGURE 6.29: Curve-fitted dependence of error bound $-log_{10}\mathcal{E}_b$ on θ_e and θ_g based on (6.6)

While the validity of these bounds can be confirmed through all previously published examples of the dynamic AMR-FDTD, the case of a dielectric waveguide directional coupler (Fig. 6.30), similar to that of [62], is also discussed here. In this example, letting $N_{AMR} = 10$, $\sigma_{AMR} = 2$ and $\theta_c = 0.7$, θ_g and θ_e were chosen, requiring that the error be less than 3%. Then, from Fig. 6.29, $\theta_e = 0.001$, $\theta_g = 0.0003$ satisfied this criterion. The excitation was a sine-wave

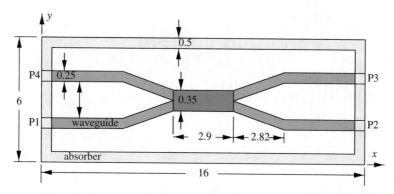

FIGURE 6.30: Geometry of a directional coupler (not drawn to scale). Dimensions are given in μm. The surrounding medium is air. The dielectric constant of the waveguide and the center piece is 3.24 and 4.41, respectively

modulated Gaussian pulse with a center frequency of 230 THz and a pulse width of 20 fs. The excitation was imposed at 1.2 μm from port 1. For AMR-FDTD, four levels of mesh resolution were used; the size of the coarsest mesh was 320×120. The time step for each mesh was determined by letting all mesh Courant numbers be equal to 0.7. Fields and errors were recorded over 3200 time steps of the root mesh. An array of 14×8 probes was used to record the E_z field throughout the computation domain. The reference result was obtained by applying FDTD in a 2560×960 uniform mesh. The actual error was 0.46%, which satisfied the requirement, although it indicated that the deduced bound was rather conservative. Finally, Fig. 6.31 compares the fields obtained by AMR-FDTD and FDTD in ports 3 and 4 of the coupler. Evidently, AMR-FDTD cannot be distinguished from the reference solution, even

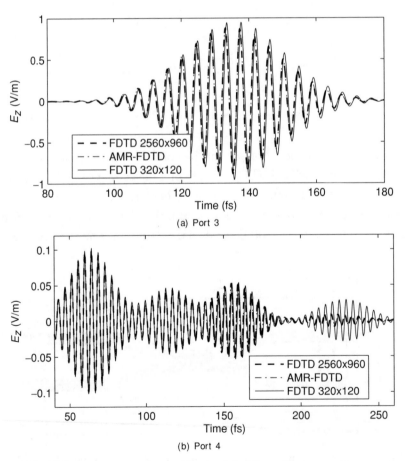

FIGURE 6.31: Electric field at port 3 and 4 of the directional coupler

for the weak field at port 4. For that waveform, the coarse mesh FDTD result suffers from a significant numerical dispersion induced error. The AMR-FDTD simulation lasts 2.8 hours, whereas the reference FDTD simulation 19.3 hours.

6.7   CONCLUSION

This chapter discussed the application of the multilevel, dynamically AMR-FDTD technique to optical waveguide structure analysis. Numerical examples demonstrated the efficiency and late-time stability of the AMR-FDTD technique. Large speed-up factors (ranging from 30 to 100) compared to the conventional FDTD were achieved, while all errors involved remained small, typically of the order of 1% or less. Furthermore, the numerical error associated with the application of the AMR-FDTD technique was studied, and its dependence on the parameters controlling the accuracy and performance of this technique was outlined. Guidelines for the *a priori* choice of these parameters were provided, thus fulfilling an important condition for AMR-FDTD to be considered as a CAD-oriented tool for electromagnetic and optical applications.

Bibliography

[1] J. B. Keller, "Geometrical theory of diffraction," *J. Opt. Soc. Am.*, vol. 52, no. 2, pp. 116–130, Feb. 1962.

[2] R. Mittra and S. W. Lee, *Analytical Techniques in the Theory of Guided Waves*. New York: Macmillan, 1971.

[3] R. F. Harrington, *Field Computation by Moment Methods*. New York: Macmillan, 1968.

[4] A. Sommerfeld, *Partial Differential Equations in Physics*. New York: Academic Press, 1964.

[5] J. Jin, *The Finite Element Method in Electromagnetics*. New York: J. W. Wiley & Sons, 1993.

[6] K. S. Yee, "Numerical solution of initial boundary value problems involving Maxwell's equations in isotropic media," *IEEE Trans. Antennas Propagat.*, vol. AP-14, pp. 302–307, Mar. 1966.

[7] W. J. R. Hoefer, "The transmission-line matrix method—Theory and applications," *IEEE Trans. Microwave Theory Tech.*, vol. MTT-33, no. 10, pp. 882–893, Oct. 1985.

[8] E. Chiprout and M. S. Nakhla, *Asymptotic Waveform Evaluation and Moment Matching for Interconnect Analysis*. Norwell, MA: Kluwer, 1994.

[9] K. L. Schlager *et al.*, "Relative accuracy of several finite-difference time-domain methods in two and three dimensions," *IEEE Trans. Antennas Propagat.*, vol. 41, no. 12, pp. 1732–1737, Dec. 1993.doi:10.1109/8.273296

[10] I. S. Kim and W. J. R. Hoefer, "A local mesh refinement algorithm for the time domain finite difference method using Maxwell's curl equations," *IEEE Trans. Microwave Theory Tech.*, vol. MTT-38, no. 6, pp. 812–815, June 1990.doi:10.1109/22.130985

[11] M. Krumpholz and L. P. B. Katehi, "MRTD: New time domain schemes based on multiresolution analysis," *IEEE Trans. Microwave Theory Tech.*, vol. MTT-44, no. 4, pp. 555–561, Apr. 1996.doi:10.1109/22.491023

[12] W. Werthen and I. Wolff, "A novel wavelet based time domain approach," *IEEE Microwave Guided Wave Lett.*, vol. 6, pp. 438–440, Dec. 1996.doi:10.1109/75.544542

[13] M. Aidam and P. Russer, "Application of biorthogonal B—spline wavelets to Telegrapher's equations," *Proc. Annu. Rev. Prog. Appl. Comput. Electromagn.*, pp. 983–990, Mar. 1998.

[14] M. Fujii and W. J. R. Hoefer, "A three-dimensional Haar wavelet-based multi-resolution analysis similar to the 3-D FDTD method—derivation and application," *IEEE Trans. Microwave Theory Tech.*, vol. MTT-46, pp. 2463–2475, Dec. 1998.

[15] S. Grivet-Talocia, "Adaptive transient solution of nonuniform multiconductor transmission lines using wavelets," *IEEE Trans. Antennas Propagat.*, vol. 48, no. 10, pp. 1563–1573, Oct. 2000.doi:10.1109/8.899673

[16] X. Zhu and L. Carin, "Multiresolution time-domain analysis of plane-wave scattering from general three-dimensional surface and subsurface dielectric targets," *IEEE Trans. Antennas Propagat.*, vol. 49, no. 11, pp. 1568–1578, Nov. 2001.doi:10.1109/8.964093

[17] E. M. Tentzeris *et al.*, "Space- and time-adaptive gridding using MRTD technique," *IEEE MTT-S Int. Microwave Symp. Dig.*, vol. 1, pp. 337–340, 1997.

[18] M. Krumpholz, E. M. Tentzeris, and L. P. B. Katehi, "Application of MRTD to printed transmission lines," *IEEE MTT-S Int. Microwave Symp. Dig.*, pp. 573–576, 1996.

[19] E. M. Tentzeris, L. Roselli, and L. P. B. Katehi, "Nonlinear circuit characterization using a multiresolution time domain technique (MRTD)," *IEEE MTT-S Int. Microwave Symp. Dig.*, vol. 3, pp. 1397–1400, 1998.

[20] E. M. Tentzeris, R. L. Robertson, and L. P. B. Katehi, "MRTD applied to complex air-dielectric boundary structures," *IEEE MTT-S Int. Microwave Symp. Dig.*, vol. 4, pp. 1463–1466, 1999.

[21] K. Goverdhanam *et al.*, "Applications of a multiresolution based FDTD multigrid," *IEEE MTT-S Int. Microwave Symp. Dig.*, vol. 1, pp. 333–336, 1997.

[22] S. G. Mallat, *A Wavelet Tour of Signal Processing*. New York: Academic Press, 1999.

[23] F. Arndt and J. Ritter, "A generalized 3D subgrid technique for the FDTD method," *IEEE MTT-S Int. Microwave Symp. Dig.*, vol. 3, pp. 1563–1566, 1997.

[24] I. Wolff and M. Walter, "An algorithm for realizing Yee's FDTD method in the wavelet domain," *IEEE MTT-S Int. Microwave Symp. Dig.*, vol. 3, pp. 1301–1304, 1999.

[25] R. G. Plumb, Z. Huang, and K. Demarest, "An FDTD/MoM hybrid technique for modeling complex antennas in the presence of heterogeneous grounds," *IEEE Trans. Geosci. Remote Sensing*, vol. 37, no. 6, pp. 2692–2698, Nov. 1999.doi:10.1109/36.803416

[26] T. D. Tsiboukis, T. I. Kosmanis, and N. V. Kantartzis, "A hybrid FDTD-wavelet-galerkin technique for the numerical analysis of field singularities inside waveguides," *IEEE Trans. Magn.*, vol. 36, no. 4, pp. 902–906, July 2000.doi:10.1109/20.877589

[27] M. Fujii *et al.*, "A 2D TLM and Haar MRTD real time hybrid connection algorithm," *Proc. Annu. Rev. Prog. Appl. Comput. Electromagn.*, pp. 1013–1020, 2000.

[28] H. Kim, S. Ju, and D.-H. Bae, "The modeling of lumped elements using the Haar wavelet multiresolution time domain technique," *Proc. IEEE AP-S Symp.*, 2000.

[29] C. D. Sarris and L. P. B. Katehi, "An efficient numerical interface between Haar MRTD and FDTD: Formulation and applications," *IEEE Trans. Microwave Theory Tech.*, vol. 51, no. 4, pp. 1146–1156, Apr. 2003.doi:10.1109/TMTT.2003.809620

[30] I. Daubechies, *Ten Lectures on Wavelets*. Philadelphia, PA: Society for Industrial and Applied Mathematics, 1992.

[31] H. L. Royden, *Real Analysis,* 3rd edn. Englewood Cliffs, NJ: Prentice Hall, 1988.

[32] G. Papanicolaou, E. Bacry, and S. Mallat, "A wavelet based space-time adaptive numerical method for partial differential equations," *Math. Modeling Numer. Anal.*, vol. 26, no. 7, p. 793, 1992.

[33] A. Haar, "Zur Theorie der Orthogonalen Funktionensysteme," *Math. Ann.*, vol. 69, pp. 331–371, 1910.doi:10.1007/BF01456326

[34] G. Battle, "A block spin construction of ondelettes, Part I : Lemarie functions," *Comm. Math. Phys.*, vol. 110, pp. 601–615, 1987.doi:10.1007/BF01205550

[35] C. Huber, M. Krumpholz, and P. Russer, "A field theoretical comparison of FDTD and TLM," *IEEE Trans. Microwave Theory Tech.*, vol. 43, pp. 1935–1950, Aug. 1995. doi:10.1109/22.402284

[36] J. C. Strikwerda, *Finite Difference Schemes and Partial Differential Equations*. Boca Raton, FL: CRC Press, 1989.

[37] T. Dogaru and L. Carin, "Application of Haar-wavelet-based multiresolution time-domain schemes to electromagnetic scattering problems," *IEEE Trans. Antennas Propagat.*, vol. 50, no. 6, pp. 774–784, June 2002.doi:10.1109/TAP.2002.1017657

[38] C. D. Sarris and L. P. B. Katehi, "Fundamental gridding related dispersion effects in MRTD schemes," *IEEE Trans. Microwave Theory Tech.*, vol. 49, no. 12, pp. 2248–2257, Dec. 2001.doi:10.1109/22.971607

[39] A. Taflove, *Computational Electrodynamics: The Finite-Difference Time-Domain Method*. Norwood, MA: Artech House, 1995.

[40] E. M. Tentzeris *et al.*, "Multiresolution time-domain (MRTD) adaptive schemes using arbitrary resolutions of wavelets," *IEEE Trans. Microwave Theory Tech.*, vol. 50, no. 2, pp. 501–516, Feb. 2002.doi:10.1109/22.982230

[41] K. Goverdhanam *et al.*, "A perfectly matched layer formulation for Haar wavelet based MRTD," *Proc. Eur. Microwave Conf.*, pp. 243–246, 1999.

[42] S. D. Gedney, "An anisotropic perfectly matched layer-absorbing medium for the truncation of FDTD lattices," *IEEE Trans. Antennas Propagat.*, vol. 44, no. 12, pp. 1360–1369, Dec. 1996.doi:10.1109/8.546249

[43] J. Citerne, G. Carat, R. Gillard, and J. Wiart, "An efficient analysis of planar microwave circuits using a DWT-based Haar MRTD scheme," *IEEE Trans. Microwave Theory Tech.*, vol. 48, no. 12, pp. 2261–2270, Dec. 2000.doi:10.1109/22.898973

[44] M. A. Stuchly, M. Okoniewski, and E. Okoniewska, "Three dimensional sub-gridding algorithm for FDTD," *IEEE Trans. Antennas Propagat.*, vol. 45, no. 3, pp. 422–429, Mar. 1997.doi:10.1109/8.558657

[45] M. Kuzyk, D. Sullivan, and J. Liu, "Three-dimensional optical pulse simulation using the FDTD method," *IEEE Trans. Microwave Theory Tech.*, vol. 48, no. 7, pp. 1127–1133, July 2000.doi:10.1109/22.848495

[46] H. G. Winful, "Pulse compression in optical fiber filters," *Appl. Phys. Lett.*, vol. 46, pp. 527–529, Mar. 1985.doi:10.1063/1.95580

[47] H. G. Winful, M. Krumpholz, and Linda P. B. Katehi, "Nonlinear time domain modeling by multiresolution time domain (MRTD)," *IEEE Trans. Microwave Theory Tech.*, vol. 45, no. 3, pp. 385–393, Mar. 1997.doi:10.1109/22.563337

[48] D. M. Sheen, S. M. Ali, M. Abouzahra, and J. A. Kong, "Application of the three-dimensional finite-difference time-domain method to the analysis of planar microstrip circuits," *IEEE Trans. Microwave Theory Tech.*, vol. 38, no. 7, pp. 849–857, July 1990. doi:10.1109/22.55775

[49] R. Luebbers, J. Schuster, and K. Wu, "Application of moving window FDTD to prediction of path loss over irregular terrain," *Proc. IEEE AP-S Symp.*, vol. 2, pp. 610–613, June 2003.

[50] R. Kastner, B. Fidel, E. Heyman, and R. W. Ziolkowski, "Hybrid ray-FDTD moving frame approach to pulse propagation," *Proc. IEEE AP-S Symp.*, vol. 3, pp. 1414–1417, 1994.

[51] R. Kastner, B. Fidel, E. Heyman, and R. W. Ziolkowski, "Absorbing boundary conditions in the context of the hybrid ray-FDTD moving window solution," *Proc. IEEE AP-S Symp.*, vol. 2, pp. 1006–1009, 1997.

[52] M. Berger and J. R. Oliger, "Adaptive mesh refinement for hyperbolic partial differential equation," *J. Computat. Phys.*, vol. 53, pp. 484–512, 1984.doi:10.1016/0021-9991(84)90073-1

[53] Y. Liu and C. D. Sarris, "Efficient modeling of microwave integrated circuit geometries via a dynamically adaptive mesh refinement (AMR)-FDTD technique," *IEEE Trans. Microwave Theory Tech.*, vol. 54, no. 2, pp. 689–703, Feb. 2006. doi:10.1109/TMTT.2005.862660

[54] G. Mur, "Absorbing boundary conditions for the finite-difference approximation of the time-domain electromagnetic field equations," *IEEE Trans. Electromagn. Comput.*, vol. EMC-23, pp. 377–382, 1981.

[55] M. Berger and I. Rigoutsos, "An algorithm for point clustering and grid generation," *IEEE Trans. Syst., Man. Cybern.*, vol. 21, no. 5, pp. 1278–1286, Sept./Oct. 1991. doi:10.1109/21.120081

[56] A. Rennings, V. Toni, P. Waldow, Y. Qian, T. Itoh, and I. Wolff, "A novel adaptivity for em time domain methods: Scale adaptive time steps (SATS)," *IEEE MTT-S Int. Microwave Symp. Dig.*, vol. 2, pp. 757–760, May 2001.

[57] T. Mangold, J. Rebel, W. J. R. Hoefer, P. P. M. So, and P. Russer, "What determines the speed of time-discrete algorithms," *Proc. Annu. Rev. Prog. Appl. Comput. Electromagn.*, vol. 2, pp. 594–601, 2000.

[58] S. S. Zivanovic, K. S. Yee, and K. K. Mei, "A subgridding method for the time-domain finite-difference method to solve Maxwells equations," *IEEE Trans. Microwave Theory Tech.*, vol. 39, no. 3, pp. 471–479, Mar. 1991.doi:10.1109/22.75289

[59] Y. Liu and C. D. Sarris, "Fast time-domain simulation of optical waveguide structures with a multilevel dynamically adaptive mesh refinement FDTD approach," *IEEE/OSA J. Lightwave Tech.*, vol. 24, no. 8, pp. 3235–3247, Aug. 2006.doi:10.1109/JLT.2006.876899

[60] M. Fujii and W. J. R. Hoefer, "Interpolating wavelet collocation method of time dependent Maxwells equations: Characterization of electrically large optical waveguide discontinuities," *J. Computat. Phys.*, vol. 186, no. 2, pp. 666–689, Apr. 2003.doi:10.1016/S0021-9991(03)00091-3

[61] S. C. Hagness, D. Rafizadeh, S. T. Ho, and A. Taflove, "FDTD microcavity simulations: Design and experimental realization of waveguide coupled single-mode ring and whispering-gallery-mode disk resonators," *IEEE/OSA J. Lightwave Tech.*, vol. 15, no. 11, pp. 2154–2165, Nov. 1997.doi:10.1109/50.641537

[62] S. T. Chu and S. K. Chaudhuri, "A finite-difference time-domain method for the design and analysis of guided-wave optical structures," *IEEE/OSA J. Lightwave Tech.*, vol. 7, no. 12, pp. 2033–2038, Dec. 1989.doi:10.1109/50.41625

Author Biography

Costas D. Sarris received a Ph.D. and a M.Sc. in Electrical Engineering, and a M.Sc. in Applied Mathematics from the University of Michigan, Ann Arbor, in 2002, 1998 and 2002, respectively. He also received a Diploma in Electrical and Computer Engineering (with distinction) from the National Technical University of Athens (NTUA), Greece, in 1997. In November 2002, he joined the Edward S. Rogers Sr. Department of Electrical and Computer Engineering (ECE), University of Toronto, Toronto, ON, Canada, where he is currently an Assistant Professor. His research interests are in the area of computational electromagnetics, with emphasis in high-order, mesh-adaptive techniques. He is currently involved with basic research in novel numerical techniques, as well as applications of time-domain analysis to wireless channel modeling, wave-propagation in complex media and meta-materials and electromagnetic compatibility/interference (EMI/EMC) problems.

Prof. Sarris has received a number of scholarship distinctions, including Hellenic Fellowship Foundation (1993–1997) and Technical Chamber of Greece (1994–1997) awards for academic excellence and an NTUA 1997 class bronze medal. He received a student paper award in the 2001 International Microwave Symposium for his work on a hybrid FDTD/MRTD numerical scheme, a Canada Foundation for Innovation New Opportunities Fund Award in 2004 and an award for excellence in undergraduate teaching from the Department of Electrical and Computer Engineering, University of Toronto.

Printed in the United States
by Baker & Taylor Publisher Services